Lecture Notes in Bioinformatic

Edited by S. Istrail, P. Pevzner, and M. Wa

Subseries of Lecture Notes in Computer Science

Mehmet M. Dalkilic   Sun Kim
Jiong Yang (Eds.)

Data Mining
and Bioinformatics

First International Workshop, VDMB 2006
Seoul, Korea, September 11, 2006
Revised Selected Papers

 Springer

Series Editors

Sorin Istrail, Brown University, Providence, RI, USA
Pavel Pevzner, University of California, San Diego, CA, USA
Michael Waterman, University of Southern California, Los Angeles, CA, USA

Volume Editors

Mehmet M. Dalkilic
Sun Kim

Indiana University
Center for Genomics and Bioinformatics
School of Informatics
Bloomington,IN 47408, USA
E-mail: {dalkilic,sunkim2}@indiana.edu

Jiong Yang
Case Western Reserve University
EECS Department, Department of Epidemiology and Biostatistics
Cleveland, OH, 44106, USA
E-mail: jiong.yang@case.edu

Library of Congress Control Number: 2006938545

CR Subject Classification (1998): H.2.8, I.5, J.3, I.2, H.3, F.1-2

LNCS Sublibrary: SL 8 – Bioinformatics

ISSN   0302-9743
ISBN-10   3-540-68970-2 Springer Berlin Heidelberg New York
ISBN-13   978-3-540-68970-6 Springer Berlin Heidelberg New York

Springer is a part of Springer Science+Business Media

springer.com

© Springer-Verlag Berlin Heidelberg 2006
Printed in Germany

Typesetting: Camera-ready by author, data conversion by Scientific Publishing Services, Chennai, India
Printed on acid-free paper   SPIN: 11960669   06/3142   5 4 3 2 1 0

Preface

This volume contains the papers presented at the inaugural workshop on Data Mining and Bioinformatics at the 32nd International Conference on Very Large Data Bases (VLDB). The purpose of this workshop was to begin bringing together researchers from database, data mining, and bioinformatics areas to help leverage respective successes in each to the others. We also hope to expose the richness, complexity, and challenges in this area that involves mining very large complex biological data that will only grow in size and complexity as genome-scale high-throughput techniques become more routine. The problems are sufficiently different enough from traditional data mining problems (outside of life sciences) that novel approaches must be taken to data mine in this area. The workshop was held in Seoul, Korea, on September 11, 2006.

We received 30 submissions in response to the call for papers. Each submission was assigned to at least three members of the Program Committee. The Program Committee discussed the submission electronically, judging them on their importance, originality, clarity, relevance, and appropriateness to the expected audience. The Program Committee selected 15 papers for presentation. These papers are in the areas of microarray data analysis, bioinformatics system and text retrieval, application of gene expression data, and sequence analysis. Because of the format of the workshop and the high number of submissions, many good papers could not be included. Complementing the contributed papers, the program of VDMB 2006 included an invited talk by Simon Mercer, Program Manager for External Research, with an empahsis on life sciences.

We would like to thank the members of the Program Committee for their hard and expert work. We would also like to thank the VLDB organizers, the external reviewers, the authors, and the participants for their contribution to the continuing success of the workshop. Thanks also to Indiana University School of Informatics for the generous financial support.

October 2006

Mehmet Dalkilic
Sun Kim
Jiong Yang
Program Chairs
VDMB 2006

VDMB 2006 Organization

Program Committee Chairs

Mehmet Dalkilic (Indiana University, USA)
Sun Kim (Indiana University, USA)
Jiong Yang (Indiana University, USA)

Program Committee Members

Mark Adams (Case Western University, USA)
Xue-wen Chen (University of Kansas, USA)
Jong Bhak (Korea Bioinformatics Center, Korea)
Dan Fay (Microsoft/Director Technical Computing, North America)
Hwan-Gue Cho (Busan National University, Korea)
Jeong-Hyeon Choi (Indiana University, USA)
Tony Hu (Drexel University, USA)
Jaewoo Kang (North Carolina State University, USA)
George Karypis (University of Minnesota, USA)
Doheon Lee (KAIST, Korea)
Jing Li (Case Western Reserve University, USA)
Yanda Li (Tsinghua University, China)
Birong Liao (Eli Lilly, USA)
Li Liao (University of Delaware, USA)
Huiqing Liu (University of Georgia, USA)
Lei Liu (University of Illinois at Urbana Champaign, USA)
Xiangjun (Frank) Liu (Tsinghua University, China)
Qingming Luo (Huazhong University, China)
Simon Mercer (Microsoft, USA)
Jian Pei (Simon Fraser University, Canada)
Meral Ozsoyoglu (Case Western Reserve University, USA)
Predrag Radivojac (Indiana University, USA)
Tetsuo Shibuya (University of Tokyo, Japan)
Keiji Takamoto (Case Western Reserve University, USA)
Haixu Tang (Indiana University, USA)
Anthony Tung (National University of Singapore, Singapore)
Wei Wang (University of North Carolina at Chapel Hill, USA)
Mohammed Zaki (Rensselaer Polytechnic Institute, USA)
Aidong Zhang (State University of New York at Buffalo, USA)

Table of Contents

Bioinformatics at Microsoft Research

Simon Mercer

Microsoft Research, One Microsoft Way, Redmond, WA 98052-6399
simon.mercer@microsoft.com

Abstract. The advancement of the life sciences in the last twenty years has been the story of increasing integration of computing with scientific research, and this trend is set to transform the practice of science in our lifetimes. Conversely, biological systems are a rich source of ideas that will transform the future of computing.

In addition to supporting academic research in the life sciences, Microsoft Research is a source of tools and technologies well suited to the needs of basic scientific research. Current projects include new languages to simplify data extraction and processing, tools for scientific workflows, and biological visualization.

Computer science researchers also bring new perspectives to problems in biology, such as the use of schema-matching techniques in merging ontologies, machine learning in vaccine design, and process algebra in understanding metabolic pathways.

M.M. Dalkilic, S. Kim, and J. Yang (Eds.): VDMB 2006, LNBI 4316, p. 1, 2006.
© Springer-Verlag Berlin Heidelberg 2006

A Novel Approach for Effective Learning of Cluster Structures with Biological Data Applications

Miyoung Shin

School of Electrical Engineering and Computer Science, Kyungpook National University,
1370 Sankyuk-dong, Buk-gu, Daegu 702-701, Korea
shinmy@knu.ac.kr

Abstract. Recently DNA microarray gene expression studies have been actively performed for mining unknown biological knowledge hidden under a large volume of gene expression data in a systematic way. In particular, the problem of finding groups of co-expressed genes or samples has been largely investigated due to its usefulness in characterizing unknown gene functions or performing more sophisticated tasks, such as modeling biological pathways. Nevertheless, there are still some difficulties in practice to identify good clusters since many clustering methods require user's arbitrary selection of the number of target clusters. In this paper we propose a novel approach to systematically identifying good candidates of cluster numbers so that we can minimize the arbitrariness in cluster generation. Our experimental results on both synthetic dataset and real gene expression dataset show the applicability and usefulness of this approach in microarray data mining.

1 Introduction

In recent years, microarray gene expression studies have been actively pursued for mining biologically significant knowledge hidden under a large volume of gene expression data accumulated by DNA microarray experiments. Particularly great attentions have been paid to data mining schemes for gene function discovery, disease diagnosis, regulatory network inference, pharmaceutical target identification, etc [1, 2, 3]. A principal task in investigating there problems is to identify gene groups or samples which show similar expression patterns over multiple experimental conditions. The detection of such co-expressed genes or samples allows us to infer their high possibility to have similar biological behaviors, so these can be used to characterize unknown biological facts as in [4,5,6,7,8,9].

So far, numerous methods have been investigated to efficiently find groups of genes or samples showing similar expression patterns. An extensive survey of clustering algorithms is given in [10]. The most widely-used algorithms for microarray data analysis are hierarchical clustering [4], k-means clustering [8], self-organizing maps [5], etc. Also, there are more sophisticated algorithms such as quantum clustering

M.M. Dalkilic, S. Kim, and J. Yang (Eds.): VDMB 2006, LNBI 4316, pp. 2–13, 2006.

with singular value decomposition [11], bagged clustering [12], diametrical clustering [13], CLICK [14], and so on.

Nevertheless, there still remain some difficulties in practice to identify good clusters in an efficient way. One of the problems is that many clustering methods require user's arbitrary selection of the number of target clusters. Moreover, the selection of the number of clusters dominates the quality of clustering results. Some recent tasks have addressed these issues for cluster analysis of gene expression data. In [15], Bolshakova *et al.* estimated the number of clusters inherent in microarray data by using the combination of several clustering and validation algorithms. On the other hand, in [16], Amato *et al.* proposed an automatic procedure to get the number of clusters present in the data as a part of a multi-step data mining framework composed of a non-linear PCA neural network for feature extraction and probabilistic principal surfaces combined with an agglomerative approach based on Negentropy. Also, a recent paper by Tseng *et al.* [17] suggested a parameterless clustering method called the correlation search technique. Yet, these methods are either still based on an arbitrary selection of the number of clusters or work only with their own clustering algorithms.

In this paper our concern is to propose a systematic approach to identify *good* number of clusters on a given data, which also can possibly work with the widely-used clustering algorithms requiring a specified number k of clusters. To realize this, we define the goodness of the number of clusters in terms of its *representational capacity* and investigate its applicability and usability in learning cluster structures with synthetic dataset and microarray gene expression dataset. The rest of the paper is organized as follows. In Section 2, we give the definition of the *representational capacity* (hereafter *RC*) and introduce its properties. Based on these, in Section 3, we present the *RC*-based algorithm for the estimation of the number of clusters on a given dataset. In Section 4, the experimental results are presented and discussed. Finally, concluding remarks are given in Section 5.

2   Definition of *RC* and Its Properties

One of the critical issues in cluster analysis is to identify the good number of clusters on a given dataset. Intuitively it may be the number of groups in which the members within the group are highly homogeneous and the members between the groups are highly separable. Without a *priori* knowledge, however, it is not easy to conjecture the good number of clusters hidden under the data in advance. To handle this issue in an efficient and systematic way, we introduce the concept of *RC* as a vehicle to quantify the goodness of the cluster number and use this to estimate the good number of clusters for given data.

In this section, the definition of *RC* and its properties are given first, and then present the algorithm to estimate the number of clusters using *RC* criterion in the following section.

2.1 Distribution Matrix

To define the *RC*, we employ the matrix which captures the underlying characteristics of given data, called the *distribution matrix*. Specifically, for the dataset $\mathbf{D} = \{\mathbf{x}_i, i = 1,\ldots,n : \mathbf{x}_i = (x_{i1},\ldots,x_{id}) \in R^d\}$, the distribution matrix $\mathbf{\Phi}$ is defined as

$$\mathbf{\Phi} = \begin{bmatrix} \phi_{11} & \phi_{12} & \cdots & \phi_{1n} \\ \phi_{21} & \phi_{22} & \cdots & \phi_{2n} \\ \vdots & \vdots & & \vdots \\ \phi_{n1} & \phi_{n2} & \cdots & \phi_{nn} \end{bmatrix} = \begin{bmatrix} \phi(\mathbf{x}_1,\mathbf{x}_1) & \phi(\mathbf{x}_1,\mathbf{x}_2) & \cdots & \phi(\mathbf{x}_1,\mathbf{x}_n) \\ \phi(\mathbf{x}_2,\mathbf{x}_1) & \phi(\mathbf{x}_2,\mathbf{x}_2) & \cdots & \phi(\mathbf{x}_2,\mathbf{x}_n) \\ \vdots & \vdots & & \vdots \\ \phi(\mathbf{x}_n,\mathbf{x}_1) & \phi(\mathbf{x}_n,\mathbf{x}_2) & \cdots & \phi(\mathbf{x}_n,\mathbf{x}_n) \end{bmatrix}$$

where $\phi_{ij} = \phi(\mathbf{x}_i,\mathbf{x}_j) = \exp(-d(\mathbf{x}_i,\mathbf{x}_j)^2/2\sigma^2)$ and $d(.)$ is a distance metric. That is, the element ϕ_{ij} reflects the normalized distance between two data vectors of ($\mathbf{x}_i$, $\mathbf{x}_j$) into the range [0, 1] by the Gaussian. Thus, the quantity of ϕ_{ij} becomes closer to 1 when $\mathbf{x}_i$ gets closer to $\mathbf{x}_j$. Conversely, ϕ_{ij} becomes closer to 0 when the vector $\mathbf{x}_i$ gets farther from $\mathbf{x}_j$. Here the closeness between the two vectors is relatively defined by the width σ of the Gaussian. For a large σ, the Gaussian has a smooth shape incurring less impact of actual distance on the quantity of ϕ_{ij}. On the other hand, for a small σ, the Gaussian has a sharp shape incurring more impact of the distance on the quantity ϕ_{ij}.

2.2 Definition of RC

Assuming that the dataset $\mathbf{D}$ has k ($< n$) generated clusters, its corresponding *RC*, denoted by $RC(\tilde{\mathbf{D}}_k)$, is defined as follows.

Definition of $RC(\tilde{\mathbf{D}}_k)$**:** For a given dataset $\mathbf{D}$ consisting of n data vectors, $RC(\tilde{\mathbf{D}}_k)$ is defined by

$$RC(\tilde{\mathbf{D}}_k) = 1 - \frac{\|\mathbf{\Phi} - \tilde{\mathbf{\Phi}}_k\|_2}{\|\mathbf{\Phi}\|_2} \quad \text{where} \quad \tilde{\mathbf{\Phi}}_k = \Sigma_{i=1}^k s_i \mathbf{u}_i \mathbf{v}_i^T \tag{1}$$

Here $\tilde{\mathbf{D}}_k$ represents the data of k generated clusters by clustering algorithm and $\mathbf{\Phi}$ is the distribution matrix of $\mathbf{D}$. For $\tilde{\mathbf{\Phi}}_k$, s_i denotes the ith singular values of $\mathbf{\Phi}$, and $\mathbf{u}_i$ and $\mathbf{v}_i$ denote the ith left and right singular vectors, respectively.

2.3 Properties of RC

The definition of $RC(\tilde{\mathbf{D}}_k)$ in Formula (1) can be also described as,

$$RC(\tilde{\mathbf{D}}_k) = 1 - \frac{s_{k+1}}{s_1} \tag{2}$$

where s_1, s_{k+1} are the 1st and the $(k+1)^{th}$ singular values, respectively, of $\mathbf{\Phi}$.

Proof. Based on Theorem 2.3.1 and Theorem 2.5.3 (see [18] for reference), the 2-norm of $\boldsymbol{\Phi}$ is the square root of the largest eigen value of $\boldsymbol{\Phi}^T\boldsymbol{\Phi}$, which is equal to the first singular value of $\boldsymbol{\Phi}$. Thus, $\left\|\boldsymbol{\Phi}\right\|_2 = s_1$. Also, $\left\|\boldsymbol{\Phi} - \tilde{\boldsymbol{\Phi}}_k\right\|_2 = s_{k+1}$, which completes the proof. $\qquad\qquad\square$

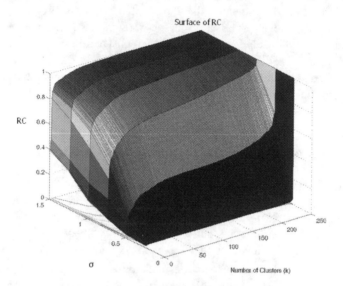

Fig. 1(a). The surface of *RC* simulated with the synthetic data for different choices of number of clusters k=(5:5:250) and the closeness parameter σ=(0.25:0.25:1.5)

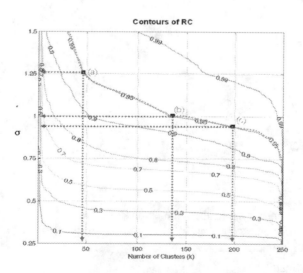

Fig. 1(b). The contours of RC simulated with the synthetic data for different choices of number of clusters k=(5:5:250) and the closeness parameter σ=(0.25:0.25:1.5)

Figures 1 (a) and (b) shows the surface and contours of *RC*, respectively, which have been simulated with synthetic data over the (k, σ) space. Here k is the number of clusters and σ is the closeness parameter. As seen in Figure 1(a), a larger number k of clusters increases the corresponding *RC* continuously up to reach a certain number of clusters and then it stays almost flat even with additional number of clusters, although the saturation point is dependent on the choice of σ. Also, in Figure 1(b), it is seen that to meet a certain level of the *RC*, a larger σ requires smaller number of clusters while a smaller σ requires larger number of clusters. Therefore, it is observed that for different choices of σ, we can have several different k's satisfying a specific RC criterion. For example, three possible choices of (k, σ) combinations denoted as (a), (b), and (c) are illustrated in the contours of Figure 1(b), where all of them have *RC* = 0.95.

3   Estimation of Number of Clusters Based on RC Criterion

In this section, we address the problem of estimating the good number of clusters for a given dataset. By using *RC*, it can be formulated as the problem of identifying the minimum number of clusters k such that its corresponding *RC* is not less than the specified *RC* having a certain amount of error allowance, say δ. Assuming an error allowance δ $(0 < \delta < 1)$, it means that we want to find "k" clusters having no less than *RC* $= 1 - \delta$. Thus, in this case, the desired number of target clusters for a given dataset **D** should be the smallest k which satisfies the following condition:

$$RC(\tilde{\mathbf{D}}_k) \geq 1 - \delta$$

By Formula (2), this can be also stated as

$$1 - \frac{s_{k+1}}{s_1} \geq 1 - \delta$$

That is,

$$\frac{s_{k+1}}{s_1} \leq \delta$$

$$s_{k+1} \leq s_1 \times \delta$$

$$s_k > s_1 \times \delta. \tag{3}$$

Now our concern is to find the smallest k satisfying condition (3). Interestingly, this problem corresponds to the problem of computing the *effective rank* (also called *ε-rank*) of the distribution matrix **Φ**. Note that for some small $\varepsilon > 0$, the *ε-rank* of a matrix $\tilde{\mathbf{D}}$ is defined as

$$r_\varepsilon = rank(\tilde{\mathbf{D}}, \varepsilon) \tag{4}$$

such that

$$s_1 \geq \cdots \geq s_{r_\varepsilon} > \varepsilon > s_{r_\varepsilon+1} \geq \cdots \geq s_n .$$

Here ε denotes the effect of noise and rounding errors in the data. Such a rank estimate r_ε is referred to as the *effective rank* of the matrix [18]. Putting the condition (3) into the *effective rank* definition shown in (4), therefore, the desirable number of clusters (k) can be obtained by estimating the ε-*rank* of Φ with taking $\varepsilon = s_1 \times \delta$.

As a consequence, for the distribution matrix Φ of a given dataset D, the minimum number k of clusters satisfying a given condition of $RC(\tilde{D}_k) \geq 1 - \delta$ can be computed by

$$k = rank(\Phi, \varepsilon) = rank(\Phi, s_1 \times \delta) .$$

4 Experimental Results

Our experiments have been performed with two datasets, a synthetic dataset and a yeast cell-cycle dataset, for both of which the true numbers of clusters are already known along with the target clusters. For our analyses, the value of a closeness parameter σ has been heuristically chosen as in the range of $0 < \sigma < \sqrt{d}/2$, where d is the dimensionality of data vectors. With different choices of error allowance δ =0.01, 0.05, 0.1, 0.2, and 0.3, we first identified the minimum number of clusters k to satisfy a given RC criterion, i.e. $1-\delta$, for the values of σ in the given range, and then generated the clusters with such a chosen k. The clustering results were then evaluated with the *adjusted rand index* as a validation index. Euclidean distance was used as a distance metric.

4.1 Experiment Methodology

4.1.1 Dataset

Synthetic data: The synthetic dataset was generated based on five different predefined time series patterns, which were partially taken from the nine synthetic time series patterns studied in [19]. This dataset includes 250 gene expression profiles consisting of their log expression measures at 10 different time points. For each of the five predefined patterns, 50 data vectors were uniformly generated by adding Gaussian noise $N(0,0.5^2)$ to it.

Yeast cell cycle data: The real dataset used for our experiments is regarding mRNA transcript levels during the cell cycle of the budding yeast *S. cerevisiae*. In [20], Cho *et al.* monitored the expression levels of 6220 genes over two cell cycles, which were collected at 17 time points taken at 10 min intervals. Out of these genes, they identified 416 genes showing the peak at different time points and categorized them into five phases of cell cycle, viz. early G1, late G1, S, G2, and M phases. Among these, by removing such genes that show the peak at more than one phase of cell cycle, 380 genes were identified and used in our experiments, whose expression levels clearly show the peak at one of the five phases of cell cycle.

4.1.2 Cluster Generation

For cluster generation, we used the seed-based clustering method which has been recently developed in [21]. The seed-based clustering method consists of two phases: seed extraction and cluster generation. The first phase of seed extraction is, given the number k of clusters, to find k good seeds of data vectors by computational analysis of given data matrix in such a way that the chosen seeds can be distinguished enough not to be very similar to each other while capturing all the unique data features (see [21] for more details). Once the seeds are chosen, the second phase proceeds to generate the clusters by using the chosen seeds as the representative vectors of potential clusters and assigning each data vector to a cluster with the closest representative vector. That is, by assigning each of the data vectors included in the dataset to the cluster of which representative vector is the most similar to the current data vector, the cluster memberships of all the data vectors are identified.

4.1.3 Cluster Assessment

Here clustering results are assessed by *adjusted rand index* (hereafter ARI), which is a statistical measure to assess the agreement between two different partitions and has been used in some previous research on gene expression data analysis [22]. The adjusted rand index is defined as in Formula (5), where a value closer to 1 implies that the two partitions are closer to perfect agreement.

Suppose that $U = \{u_1,\ldots,u_R\}$ is the true partition and $V = \{v_1,\ldots,v_C\}$ is a clustering result. Then, according to [6], the adjusted rand index is defined as follows:

$$\frac{\sum_{i,j}\binom{n_{ij}}{2}-\left[\sum_i\binom{n_{i.}}{2}\sum_j\binom{n_{.j}}{2}\right]/\binom{n}{2}}{\frac{1}{2}\left[\sum_i\binom{n_{i.}}{2}+\sum_j\binom{n_{.j}}{2}\right]-\left[\sum_i\binom{n_{i.}}{2}\sum_j\binom{n_{.j}}{2}\right]/\binom{n}{2}} \tag{5}$$

where n is the total number of genes in the dataset, n_{ij} is the number of genes that are in both class u_i and cluster v_j, and $n_{i.}$ and n_j are the number of genes in class u_i and cluster v_j, respectively.

4.2 Analysis Results on Synthetic Data

For the synthetic data, we chose the closeness parameter σ in the range of $\sigma=(0.25:0.25:1.5)$. Recall that the value of σ is heuristically determined in the range of $0< \sigma < \sqrt{d}/2$, where d is the dimensionality of data vectors. Since the number of conditions in the synthetic data is 10, the range of σ was chosen as $0< \sigma < \sqrt{10}/2$, that is 1.581. Table 1 shows numerically the *RC*-based automatically chosen number of

Table 1. *RC*-based automatically chosen number of clusters (*k*) for the synthetic data with different choices of error allowance δ =0.01, 0.05, 0.1, 0.2, and 0.3. Only the dark ones are considered for cluster evaluation.

	δ =0.01	δ =0.05	δ =0.1	δ =0.2	δ =0.3
σ =0.25	250	250	250	250	250
σ =0.5	250	250	250	250	250
σ =0.75	250	248	224	104	30
σ =1.0	244	133	53	17	7
σ =1.25	170	48	25	6	5
σ =1.5	100	30	9	5	4

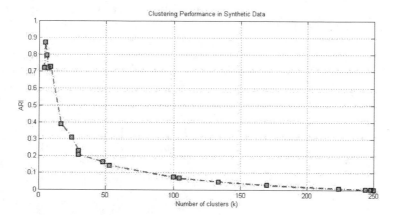

Fig. 2(a). A graphical description of RC-based automatically chosen number of clusters and their corresponding clustering performance with the synthetic data shown in Table 1. The marked squares denote the RC-based chosen numbers of clusters by our algorithm, and only these numbers of clusters are evaluated for identifying the best partition in the synthetic data.

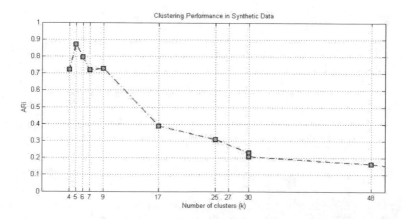

Fig. 2(b). A zoomed-in graph of Figure 2(a) with the emphasis on the RC-based automatically chosen optimal number (k=5) and its near-by number of clusters and their corresponding clustering performance with the synthetic data

clusters (k) and their corresponding clustering performance on the synthetic data for different choices of error allowance δ =0.01, 0.05, 0.1, 0.2, and 0.3. Figure 2(a) shows these results graphically all together. Out of these, as shown in Figure 2(b), the best clustering results ARI=0.869 are obtained at k=5, which is known as the *real* optimal number of clusters. Therefore, it confirms the usability of the *RC*-based chosen number of clusters in the synthetic data analysis.

4.3 Analysis Results on Yeast Cell Cycle Data

For the yeast cell cycle data, we used the closeness parameter σ in the range of σ = (0.25:0.25:2.0) because the number of conditions in this data is 17 and so the range of σ was chosen as $0 < \sigma < \sqrt{17/2}$, viz. 2.0616. Table 2 shows numerically the *RC*-based automatically chosen number of clusters (k) and their corresponding clustering performance on the yeast cell cycle data for different choices of error allowance δ =0.01,

Table 2. *RC*-based automatically chosen number of clusters (k) and their corresponding clustering performance with the yeast cell cycle data for different choices of error allowance δ =0.01, 0.05, 0.1, 0.2, and 0.3

	δ =0.01	δ =0.05	δ =0.1	δ =0.2	δ =0.3
σ =0.25	384	384	384	384	384
σ =0.5	384	384	384	384	384
σ =0.75	384	384	381	333	20
σ =1.0	384	346	121	14	5
σ =1.25	368	149	27	8	4
σ =1.5	305	50	18	7	3
σ =1.75	210	32	13	6	4
σ =2.0	138	23	11	5	3

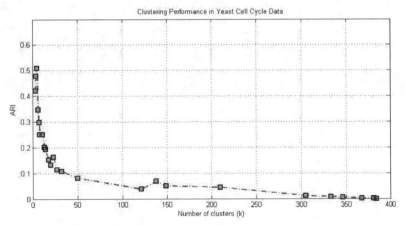

Fig. 3(a). A graphical description of *RC*-based automatically chosen number of clusters and their corresponding clustering performance with the yeast cell cycle data shown in Table 2. The marked squares denote the *RC*-based chosen numbers of clusters by our algorithm, and only these numbers of clusters are evaluated for identifying the best clustering results in the yeast cell cycle data.

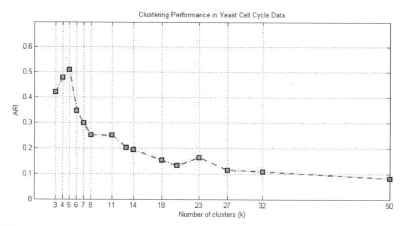

Fig. 3(b). A zoomed-in graph of Figure 3(a) with the emphasis on the *RC*-based automatically chosen *optimal* number (*k*=5) and its near-by number of clusters and their corresponding clustering performance with the yeast cell cycle data

0.05, 0.1, 0.2, and 0.3. Figure 3(a) shows these results graphically all together. Out of these, as shown in Figure 3(b), the best clustering result of ARI=0.509 is obtained at *k*=5, which is known as the *real* optimal number of clusters in the biological sense. Thus, it confirms the usability of the *RC*-based automatically chosen number of clusters in the yeast cell cycle data analysis.

5 Discussions and Concluding Remarks

In this paper we addressed the problem of estimating the good number of clusters for given data based on a newly introduced measure called *RC* criterion. Since the quantity of *RC* has such characteristics that it increases with the number of clusters up to a certain point and then staying flat after the point, the saturation point may be the good number of clusters. By taking an error allowance δ for *RC*, possible candidates of the good number of clusters for given data are systematically identified for different choices of the closeness parameter. The value of the closeness parameter is recommended to be taken properly up to the square root of the data dimension divided by two with a reasonable interval. A dense interval would lead to more detailed analyses while the coarse interval to rough analyses. As a consequence, the proposed method can save computational costs by trying a fewer values of *k*'s to identify the best partition with any clustering algorithms requiring a specified *k*. In an unsupervised setting, cluster validation should be done with internal indices, such as Silhouette index, Dunn's index, and etc, instead of the adjusted rand index.

Our experimental results on synthetic data and real biological data showed the high potential in usability and applicability of the proposed method to cluster structure learning tasks. Further, it was also shown that this method is favorable for gene expression data analysis, as an alternative to current arbitrary selection methods. In spite of additional costs for computing *RC* itself, we believe that, due to its very attractive features, the *RC*-based estimation of the number of clusters should be very appealing to practitioners in terms of the easiness of use and performance. For more extensive applications, however, other distance metrics, such as Pearson correlation coefficients, need to be investigated, which would be our future works.

References

1. D. J. Hand and N. A. Heard, Finding groups in gene expression data, Journal of Biomedicine and Biotechnology (2005). Vol. 2, 215-225.
2. D. K. Slonim, From patterns to pathways: gene expression data analysis comes of age, Nature genetics supplement (2002), vol. 32, 502-508.
3. M.G. Walker, Pharmaceutical target identification by gene expression analysis, Mini reviews in medicinal chemistry (2001), Vol. 1, 197-205.
4. M.B. Eisen, P.T. Spellman, P.O. Brown and D. Bostein: Cluster analysis and display of genome-wide expression patterns, Proc. Natl. Acad. Sci. (1998), Vol. 95, 14863-14868.
5. P. Tamayo et al., Interpreting patterns of gene expression with self-organizing maps: methods and application to hematopoietic differentiation, Proc. Natl. Acad. Sci. (1999), Vol. 96, 2907-2912.
6. T.R. Golub et al., Molecular classification of cancer: class discovery and class prediction by gene expression monitoring, Science (1999), Vol. 286, 531-537.
7. H. Liu, J. Li and L. Wong, Use of extreme patient samples for outcome prediction from gene expression data, Bioinformatics (2005), Vol. 21(16), 3377-3384.
8. S. Tavazoie, J.D. Hughes, M.J. Campbell, R.J. Cho and G. M. Church: Systematic determination of genetic network architecture, Nature Genetics (1999), Vol. 22, 281-285.
9. H. Toh and K. Horimoto, Inference of a genetic network by a combined approach of cluster analysis and graphical Gaussian modeling, Bioinformatics (2002), Vol. 18(2), 287-297
10. R.Xu and D. Wunsch II, Survey of clustering algorithms, IEEE Trans. on Neural Networks (2005), Vol. 16(3), 645-678.
11. D. Horn and I. Axel, Novel clustering algorithm for microarray expression data in a truncated SVD space, Bioinformatics (2003), Vol. 19, 1110-1115.
12. S. Dudoit and J. Fridlyand, Bagging to improve the accuracy of a clustering procedure, Bioinformatics (2003), Vol. 19, 1090-1099.
13. I. Dhilon et al. Diametrical clustering for identifying anti-correlated gene clusters, Bioinformatics, Vol. 19, 1612-1619.
14. R. Sharan et al., Click and expander: a system for clustering and visualizing gene expression data, Bioinformatics (2003). Vol. 19, 1787-1799.
15. N. Bolshakova and F. Azuaje, Estimating the number of clusters in DNA microarray data, Methods Inf. Med (2006), Vol. 45(2), 153-157.
16. R. Amato et al., A multi-step approach to time series analysis and gene expression clustering, Bioinformatics (2006), Vol. 22(5), 589-596.
17. V. S. Tseng and C-P. Kao, Efficiently mining gene expression data via a novel parameterless clustering method, IEEE/ACM trans. on Comp. Biology and Bioinformatics (2005), Vol. 2(4), 355-365.
18. G.H. Golub and C.F. Van Loan, Matrix Computation (3rd edition), The Johns Hopkins University Press (1996)
19. J. Quackenbush, Computational analysis of microarray data, Nature Reviews Genetics (2001), Vol. 2, 418-422.

20. R.J. Cho et al., A genome-wide transcriptional analysis of the mitotic cell cycle, Molecular Cell, 2:65-73, 1998.
21. M. Shin and S.H. Park, Microarray expression data analysis using seed-based clustering method, Key engineering materials (2005), Vol. 277, 343-348
22. K.Y. Yeung, D.R. Haynor and W. L. Ruzzo: Validating clustering for gene expression data, Bioinformatics (2001), Vol. 17(4), 309-318.

Subspace Clustering of Microarray Data Based on Domain Transformation

Jongeun Jun[1], Seokkyung Chung[2,*], and Dennis McLeod[1]

[1] Department of Computer Science, University of Southern California,
Los Angeles, CA 90089, USA
[2] Yahoo! Inc., 2821 Mission College Blvd, Santa Clara, CA 95054, USA
jongeunj@usc.edu, schung@yahoo-inc.com, mcleod@usc.edu

Abstract. We propose a mining framework that supports the identification of useful knowledge based on data clustering. With the recent advancement of microarray technologies, we focus our attention on gene expression datasets mining. In particular, given that genes are often co-expressed under subsets of experimental conditions, we present a novel subspace clustering algorithm. In contrast to previous approaches, our method is based on the observation that the number of subspace clusters is related with the number of maximal subspace clusters to which any gene pair can belong. By performing discretization to gene expression profiles, the similarity between two genes is transformed as a sequence of symbols that represents the maximal subspace cluster for the gene pair. This domain transformation (from genes into gene-gene relations) allows us to make the number of possible subspace clusters dependent on the number of genes. Based on the symbolic representations of genes, we present an efficient subspace clustering algorithm that is scalable to the number of dimensions. In addition, the running time can be drastically reduced by utilizing inverted index and pruning non-interesting subspaces. Experimental results indicate that the proposed method efficiently identifies co-expressed gene subspace clusters for a yeast cell cycle dataset.

1 Introduction

With the recent advancement of DNA microarray technologies, the expression levels of thousands of genes can be measured simultaneously. The obtained data are usually organized as a matrix (also known as a gene expression profile), which consists of m columns and n rows. The rows represent genes (usually genes of the whole genome), and the columns correspond to the samples (e.g. various tissues, experimental conditions, or time points).

Given this rich amount of gene expression data, the goal of microarray analysis is to extract hidden knowledge (e.g., similarity or dependency between genes) from this matrix. The analysis of gene expressions may identify mechanisms of

---

* To whom correspondence should be addressed. Part of this research was conducted when the author was at University of Southern California.

M.M. Dalkilic, S. Kim, and J. Yang (Eds.): VDMB 2006, LNBI 4316, pp. 14–28, 2006.
© Springer-Verlag Berlin Heidelberg 2006

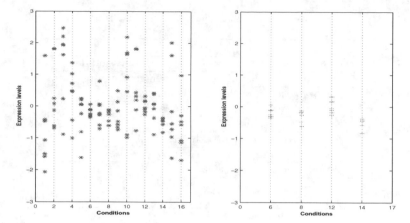

Fig. 1. Plot of sample gene expression data across whole conditions (shown in left hand side) versus subset of conditions (shown in right hand side)

gene regulation and interaction, which can be used to understand a function of a cell [6]. Moreover, comparison between expressions in a diseased tissue and a normal tissue will further enhance our understanding in the disease pathology [7]. Therefore, data mining, which transforms a raw dataset into useful higher-level knowledge, becomes a must in bioinformatics.

One of the key steps in gene expression analysis is to perform clustering genes that show similar patterns. By identifying a set of gene clusters, we can hypothesize that the genes clustered together tend to be functionally related. Traditional clustering algorithms have been designed to identify clusters in the full dimensional space rather than subsets of dimensions [6,14,4,11]. When correlations among genes are not apparently visible across the whole dimensions as shown in the left-side graph of Figure 1, the traditional approaches fail to detect clusters. However, it is well-known that genes can manifest a coherent pattern under subsets of experimental conditions as shown in the right-side graph of Figure 1. Therefore, it is essential to identify such local patterns in microarray datasets, which is a key to revealing biologically meaningful clusters.

In this paper, we propose a mining framework that supports the identification of meaningful subspace clusters. When m (the number of dimensions) is equal to 50 (m for gene expression data usually varies from 20 to 100), the number of possible subspaces is $2^{50} - 1 \approx 1.1259 \times 10^{15}$. Thus, it is computationally expensive to search all subspaces to identify clusters. To cope with this problem, many subspace clustering algorithms first identify clusters in low dimensional spaces, and use them to find clusters in higher dimensional spaces based on *apriori* principle [1] [1]. However, this approach is not scalable to the number of dimensions in general.

In contrast to the previous approaches, our method is based on the observation that the maximum number of subspaces is related with the number of maximal

---

[1] If a collection of points C form a dense cluster in a k-dimensional space, then any subset of C should form a dense cluster in a $(k-1)$-dimensional space.

Table 1. Notations for subspace clustering

Notation	Meaning
n	The total number of genes in gene expression data
m	The total number of dimensions in gene expression data
X	$n \times m$ gene expression profile matrix
x_i	The i-th gene
x_{ij}	The value of x_i in j-th dimension
S_i	The set of symbols for i-th dimension ($S_i \cap S_j = \emptyset$ if $i \neq j$)
s_{ij}	A symbol for x_{ij}
s_i	A sequence of symbols for x_i
K	The maximum number of symbols
p_i	A pattern
mp_{ij}	The maximal pattern for x_i and x_j
P_a	A set of maximal patterns that contains a symbol a
P_i	The set of all maximal patterns in i-dimensional subspace
P	The set of all maximal patterns in X
α_i	The minimum number of elements that an i-dimensional subspace cluster should have
β	The minimum dimension of subspace clusters

subspace clusters to which any two genes can belong. In particular, by performing discretization to gene expression profiles, we can transform the similarity between two genes as a sequence of symbols that represent the maximal subspace cluster for the gene pairs. This transformation allows us to limit the number of possible subspace clusters to $\frac{n(n-1)}{2}$ where n is the number of genes. Based on the transformed data, we present an efficient subspace clustering algorithm that is scalable to the number of dimensions. Moreover, by utilizing inverted index and pruning (even conservatively) non-interesting subspace clusters, the running time can be drastically reduced. Note that the presented clustering algorithm can detect subspace clusters regardless of whether their coordinate values are consecutive to each other or not.

The remainder of this paper is organized as follows. In Section 2, we explain the proposed subspace clustering algorithm. Section 3 presents experimental results. In Section 4, we briefly review the related work. Finally, we conclude the paper and provide our future plans in Section 5.

2   The Subspace Clustering Algorithm

In this section, we present the details of each step of our approach. Table 1 shows the notations, which will be used throughout this paper.

The first step of our clustering is to quantize gene expression dataset (Section 2.1). The primary reason for discretization is that we need to efficiently extract a maximal subspace cluster to which every gene pair belongs. In-depth discussions on why we need discretization, and details of our algorithm are presented in Section 2.2. In Section 2.3, we explain how we can further reduce computational cost by utilizing inverted index and pruning. Finally, in Section 2.4, we explain how to select meaningful subspace clusters. Figure 2 sketches the proposed subspace clustering algorithm.

2.1 Discretization

In general, discretization approach can be categorized into three ways:

- *Equi-width bins*. Each bin has approximately same size.
- *Equi-depth bins*. Each bin has approximately same number of data elements.
- *Homogeneity-based bins*. The data elements in each bin are similar to each other.

In this paper, we use a homogeneity-based bins approach. In particular, we utilize K-means clustering with Euclidean distance metric [5]. Because we apply K-means to 1-dimensional data, each cluster corresponds to the interval. Additionally, K corresponds to the number of symbols for discretization.

Once we identify clusters, genes belonging to a same cluster are discretized with a same symbol. That is, x_{id} and x_{jd} are represented as a same symbol if and only if $x_{id}, x_{jd} \in C_{kd}$ where C_{kd} is k-th cluster in d-th dimension. The complexity for this step is $O(nmK)$ where n and m correspond to the number of genes and conditions, respectively. In this paper, we use same value of K across all dimensions for simplicity.[2] However, in Section 3, we investigate the effect of different value of K in terms of running time.

2.2 Identification of Candidate Subspace Clusters Based on Domain Transformation

The core of our approach lies in domain transformation to tackle subspace clustering. That is, the problem of subspace clustering is significantly simplified by transforming gene expression data into domain of gene-gene relations (Figure 3). In this Section, we explore how to identify candidate subspace clusters based on the notion of domain transformation. Before we present detailed discussions on the proposed clustering algorithm, definitions for basic terminology are provided first.

Definition 1 (Induced-string). *A string $string_1$ is referred to as an induced-string of $string_2$ if every symbol of $string_1$ appears in $string_2$ and length of $string_1$ is less than $string_2$.*

Definition 2 ((Maximal) pattern). *Given s_i and s_j, which are symbolic representations of x_i and x_j, respectively, any induced-string of both s_i and s_j is referred to as a pattern of x_i and x_j. A pattern is referred to as a maximal pattern of x_i and x_j (denoted as mp_{ij}) if and only if every pattern of x_i and x_j (except mp_{ij}) is an induced-string of mp_{ij}.*

---

[2] More effective discretization algorithm for gene expression data is currently under development.

Step 1. Discretization and symbolization of gene expression data:
   Initialization: $i = 1$.
   1.1 Perform K-means clustering on i-th dimension.
   1.2 Using S_i, assign symbols to each cluster.
   1.3 $i = i + 1$.
   Repeat Step 1 for $i \leq m$.

Step 2. Enumeration of all possible maximal patterns for each gene pair
          and construction of inverted index for each dimension and Hash Table:
   2.1 Extract a maximal pattern (mp_{ij}) between x_i and x_j using s_i and s_j.
   2.2 Insert mp_{ij} to inverted index at dimension k if the length of
        mp_{ij} is equal to k.
   2.3 Insert the gene pair x_i and x_j to Hash Table indexed by mp_{ij}.
   Repeat Step 2 for all gene pairs.

Step 3. Computation of support, pruning and generating candidate clusters:
   Initialization: $i = \beta$ where β is a user-defined threshold for the
                   minimum dimension of a subspace cluster.
   3.1 Compute support of each maximal pattern (p) at i using inverted index.
   3.2 Prune p at i and super-pattern of p at $i + 1$ if support of p is less than α_i.
   3.3 $i = i + 1$
   Repeat Step 3 for $i \leq m - 1$.

Step 4. Identification of interesting subspace clusters:
   Initialization: $i = \beta$.
   4.1 For each pattern (p) of candidate clusters at i, calculate difference between
        its support and support of each super-pattern of p at $i + 1$, and assign
        the biggest difference as maximum difference of p.
   4.2 Remove p from candidate clusters at i if maximum difference of p is less
        than user-defined threshold.
   4.3 $i = i + 1$
   Repeat Step 4 for $i \leq m - 1$.

Fig. 2. An overview of the proposed clustering algorithm

For example, if $s_1 = a_1 b_1 c_2 d_3$ and $s_2 = a_1 b_1 c_1 d_3$, then maximal pattern for x_1 and x_2 is $a_1 b_1 d_3$. Since the length of the maximal pattern is 3, the maximum dimensions of candidate subspace cluster that can host x_1 and x_2 is 3. Thus, the set of genes that have same maximal pattern can be a candidate maximal subspace cluster. By discretizing gene expression data, we can obtain the upper bound for the number of maximal patterns, which is equal to $\frac{n(n-1)}{2}$.

Because a pattern is represented as a form of string, based on the definition of induced-string, the notion of super-pattern is defined as follows:

Definition 3 (Super-pattern). *A pattern p_i is a super-pattern of p_j if and only if p_j is an induced string of p_i.*

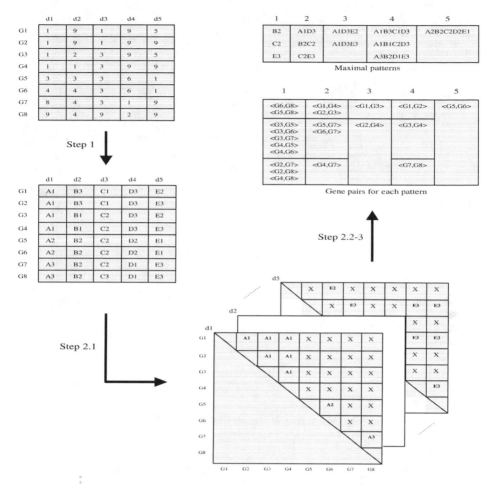

Fig. 3. An illustration of domain transformation

Figure 4 illustrates how we approach the problem of subspace clustering. Each vertex represents a gene, and the edge between vertexes shows a maximal pattern between two genes. In order to find meaningful subspace clusters, we define the minimum number of objects (α_i) that an i-dimensional subspace cluster should have. For this purpose, the notion of support is defined as follows:

Definition 4 (Support). *Given a pattern p_k, support of p_k is the total number of gene pairs (x_i and x_j) such that p_k is an induced-string of mp_{ij} or p_k is a maximal pattern of x_i and x_j.*

Thus, the problem of subspace clustering is reduced to computing support of each maximal pattern, and identifying maximal patterns whose support exceeds α_i if the length of maximal pattern is equal to i. The maximal pattern between x_1 and x_3 is defined as a_1b_1 in Figure 4. This means that x_1 and x_3 has potential to be grouped together in 2-dimensional subspace. However, we need to consider

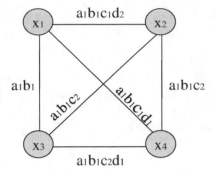

	2	3	4
a_1b_1	<X₁ X₃>	<X₂ X₃> <X₂ X₄>	< X₃ X₄> <X₁ X₄> <X₁ X₂>
$a_1b_1c_2$		<X₂ X₃> <X₂ X₄>	<X₃ X₄>
$a_1b_1c_1d_2$			<X₁ X₂>
$a_1b_1c_1d_1$			<X₁ X₄>
$a_1b_1c_2d_1$			<X₃ X₄>

Fig. 4. A sample example of subspace cluster discovery

whether there exists enough number of gene pairs that has a pattern a_1b_1. This step can be achieved by computing support for a_1b_1, and checking whether the support exceeds α_2 or not. In this example, support for a_1b_1 is 6.

2.3   Efficient Super-Pattern Search Based on Inverted Index

Achieving an efficient super-pattern search for support computation is important in our algorithm. The simple approach to searching super-patterns for a given maximal pattern at dimension d is to conduct a sequential scan on all maximal patterns at dimension d' $(d' > d)$. The following shows the running time for this approach.

$$\sum_{i=\beta}^{n-1} |P_i| \sum_{j=i+1}^{m} (j \times |P_j|) \tag{1}$$

where $|P_j|$ is the number of maximal patterns at dimension j and β is the minimum dimension of subspace clusters.

However, we can reduce the time by utilizing inverted index, which has been widely used in modern information retrieval. In inverted index [10], the index associates a set of documents with terms. That is, for each term t_i, we build a document list (D_i) that contains all documents containing t_i. Thus, when a query q is composed of $t_1, ..., t_k$, to identify a set of documents which contains q, it is

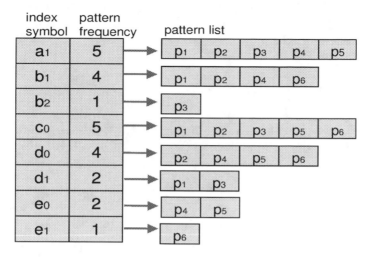

$$\mathbf{p}_1 = a_1 b_1 c_0 d_1 \; ; \mathbf{p}_2 = a_1 b_1 c_0 d_0 \; ; \mathbf{p}_3 = a_1 b_2 c_0 d_1$$
$$\mathbf{p}_4 = a_1 b_1 d_0 e_0 \; ; \mathbf{p}_5 = a_1 c_0 d_0 e_0 \; ; \mathbf{p}_6 = b_1 c_0 d_0 e_1$$

Fig. 5. Inverted index at 4-dimensional subspace

sufficient to examine the documents that contain all t_i's (i.e., intersection of D_i's) instead of checking whole document dataset. By implementing each document list as a hash table, looking up documents that contain t_i takes constant time.

In our super-pattern search problem, each term and document correspond to symbol and pattern, respectively. Figure 5 illustrates the idea. Each pattern list is also implemented as a hash table. When a query pattern is composed of multiple symbols, symbols are sorted (in ascending order) according to their pattern frequency. After then, we start to take intersection of pattern lists whose pattern frequency is lowest. For example, given a query pattern $b_2 c_0 d_0$, we search super pattern as follows: [3]

$$(P_{b_2} \cap P_{d_0}) \cap P_{c_0} = (\{p_3\} \cap \{p_2, p_4, p_5, p_6\}) \cap P_{c_0} = \emptyset$$

where P_a is a pattern list for a symbol a.

By taking intersection of pattern list whose size is small, we can reduce the number of operations. The worst time complexity for the identification of all super-patterns for all query patterns $(a = (a_1, a_2..., a_k))$ in P_k is shown as follows:

$$T_k = \sum_{i=k+1}^{m} \sum_{a \in P_k} (min(|P_{a_1}^i|, ..., |P_{a_k}^i|) \times (k-1)) \tag{2}$$

where $|P_{a_k}^i|$ corresponds to the length of pattern list for the symbol a_k at dimension i.

---

[3] If we reach an empty set while taking intersection, then there is no need to keep intersection of pattern lists.

Note that the worst time complexity for the construction of inverted index at each dimension k takes $k \times |P_k|$. Thus, the total time complexity for super-pattern search is given as below:

$$\sum_{k=\beta}^{m}(k \times |P_k|) + \sum_{k=\beta}^{m-1}(klogk + |P_k| \times T_k) \tag{3}$$

Moreover, we can further reduce running time through the pruning step. In particular, we use the notion of expected support in Zaki *et al.* [15]. That is, if support for p_{ij} in dimension k is less than α_k, then besides eliminating p_{ij} in future consideration of subspace clustering, we do not consider super-patterns of p_{ij} in dimension $k + 1$ any more.

2.4   Identification of Meaningful Subspace Clusters

After pruning maximal patterns using the notion of support, the final step is to identify *meaningful* subspace clusters, which corresponds to the Step 4 in Figure 2. Since we are interested in maximal subspace clusters, we scan maximal patterns (denoted as p) at dimension i, and compare support of p with support of all super-patterns (denoted as p') at dimension $(i + 1)$. Any maximal pattern at i (p) are not considered for a subspace cluster if the difference between support of p and that of p' is less than user-defined threshold.

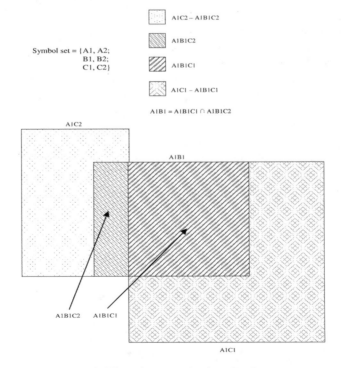

Fig. 6. Identification of subspace clusters

We illustrate this step with Figure 6. As shown, there are three maximal patterns at dimension 2 (A1B1, A1C1 and A1C2) and two maximal patterns at dimension 3 (A1B1C1 and A1B1C2), respectively. The cluster with a pattern A1B1 (C_1) contains the clusters with a pattern A1B1C1 (C_2) and A1B1C2 (C_3). If the support between C_1 and C_2 (or the support between C_1 and C_3) is less than user-defined threshold, then C_1 is ignored and both C_2 and C_3 are kept as meaningful subspace clusters. Otherwise, C_1 is also considered as a meaningful cluster besides C_2 and C_3.

3   Experimental Results

In this section, we present experimental results that demonstrate the effectiveness of the proposed clustering algorithm. Section 3.1 illustrates our experimental setup. Experimental results are presented in Section 3.2.

3.1   Experimental Setup

For empirical evaluation, the proposed clustering algorithm was tested on yeast cell cycle data. The data is a subset from the original 6,220 genes with 17 time points listed by [3]. Cho *et al.* sampled 17 time points at 10 minutes time interval, covering nearly two full cell cycles of yeast *Saccharomyces cerevisiae*. Among 6,220 genes, 2,321 genes were selected based on the largest variance in their expression. In addition, one abnormal time point was removed from the data set as suggested by [11], consequently, the resulting data consists of 2,321 genes with 16 time points. Our analysis is primarily focused on this data set.

3.2   Experimental Results

Figure 7 demonstrates how well our clustering algorithm was able to capture highly correlated clusters under a subset of conditions. The x-axis represents the conditions, and the y-axis represents expression level. As shown, subspace clusters do not necessarily manifest high correlations across all conditions. We also verified biological meanings of subspace clusters with the Gene Ontology [2]. For example, YBR279w and YLR068w (Figure 7(b)) have the most specific common parent "nucleobase, nucleoside, nucleotide and nucleic acid metabolism" in the Gene Ontology. However, they showed low correlation in full dimensions. In addition, YBR083w and YDL197C (Figure 7(d)) belonging to subspace cluster 625 participate in a similar biological function, "transcriptional control".

In order to investigate the number of maximal patterns in various scenarios, pattern rate, which represents how many maximal patterns are generated in comparison with the number of possible maximal patterns, is defined as follows:

$$Pattern\ Rate = \#maximal\ patterns\ /\ \frac{n(n-1)}{2} \qquad (4)$$

Figure 8 shows relationships between the number of genes and pattern rate. The x-axis represents the number of genes (n), and the y-axis represents the

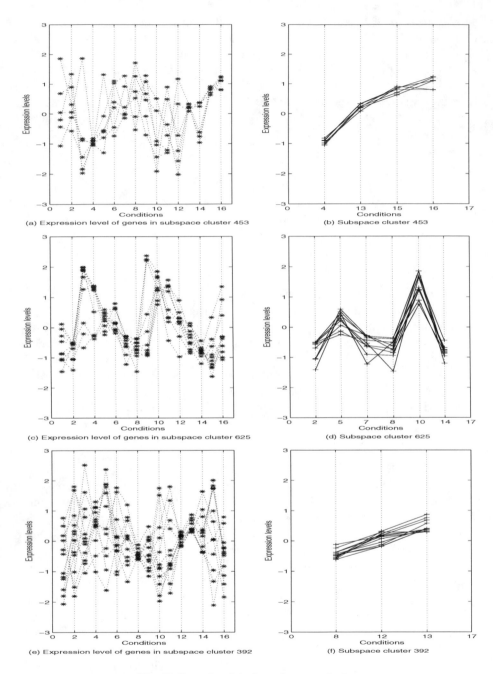

Fig. 7. Sample plots for subspace clusters

pattern rate. As shown, pattern rate decreases as n increases. This property is related with the nature of datasets. In general, datasets should have certain amount of correlations among object pairs. Otherwise, clustering cannot detect

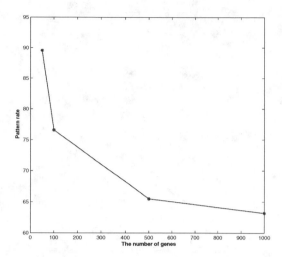

Fig. 8. The number of genes versus pattern rate

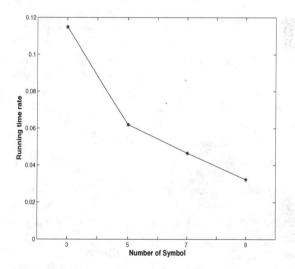

Fig. 9. Different number of symbols versus running time rate

meaningful clusters from the dataset. Thus, the increasing rate for the number of actual maximal patterns is much less than the increasing rate for possible number of maximal patterns ($\frac{n(n-1)}{2}$).

Figure 9 shows how much running time can be improved by using inverted index. The x-axis represents the number of symbols, and the y-axis represents the ratio of Equation (3) to Equation (1), which is defined as running time rate. As shown, we could observe significant improvement. For example, when the number of symbols is 7. the running time was reduced more than 20 times. Moreover, we could observe performance improvement as the number of symbols

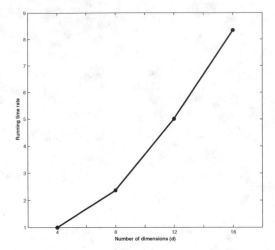

Fig. 10. The number of dimensions versus running time rate with the use of inverted index

increases. This is primarily because the length of pattern list in inverted index decreases as the number of symbols increases.

Figure 10 shows the scalability of our algorithm in terms of the number of dimensions. The x-axis represents the number of dimensions, and the y-axis represents running time rate, which is defined by running time at d divided by running time when d is 4. As shown, we could observe linear increase in running time rate, which supports the scalability of our algorithm.

4   Related Work

In this section, we briefly discuss key differences between our work and previous approaches on subspace clustering. Jiang *et al.* [8] and Parsons *et al.* [9] provide the comprehensive review on gene expression clustering and subspace clustering, respectively. For details, refer to those paper [8,9].

Recently, clustering on the subset of conditions has received significant attentions during the past few years [1,12,13,15]. Many approaches first identify clusters in low dimensions (C) and derive clusters high dimensions based on C using the *apriori* principle. However, our method is different from other approaches in that all maximal subspaces for gene pairs (whose maximum value is bounded by $n(n-1)/2$) are generated by transforming a set of genes into a set of gene-gene relations. Identification of meaningful subspace clusters is based on the maximal patterns for gene pairs.

5   Conclusion and Future Work

We presented the subspace mining framework that is vital to microarray data analysis. An experimental prototype system has been developed, implemented, and tested to demonstrate the effectiveness of the proposed algorithm. The uniqueness of our approach is based on observation that the maximum number of subspaces is limited to the number of genes. By transforming a gene-gene relation into a maximal pattern that is represented as a sequence of symbols, we could identify subspace clusters very efficiently by utilizing inverted index and pruning non-interesting subspaces. Experimental results indicated that our method is scalable to the number of dimensions.

We intend to extend this work into the following three directions. First, we are currently investigating efficient discretization algorithm for gene expression data. Second, we plan to conduct more comprehensive experiments on diverse microarray datasets as well as compare our approach with other competitive algorithms. Finally, in order to interpret obtained gene clusters, external knowledge needs to be involved. Toward this end, we plan to explore the methodology to quantify relationship between subspace clusters and known biology knowledge by utilizing Gene Ontology [2].

Acknowledgement

We would like to thank Dr. A. Fazel Famili at National Research Council of Canada for providing Cho's data. This research has been funded in part by the Integrated Media Systems Center, a National Science Foundation Engineering Research Center, Cooperative Agreement No. EEC-9529152.

References

1. R. Agrawal, J. Gehrke, D. Gunopulos and P. Raghavan. Automatic subspace clustering of high dimensional data for data mining applications. In *Proceedings of ACM SIGMOD International Conference on Management of Data*, 1998.
2. The Gene Ontology Consortium. Creating the gene ontology resource: design and implementation. *Genome Research*, 11(8):1425-1433, 2001.
3. R.J. Cho, M.J. Campbell, E.A. Winzeler,L. Steinmetz, A. Conway, L. Wodicka, T.G. Wolfsberg, A.E. Gabrielian, D. Landsman, D.J. Lockhart, and R.W. Davis. A genome-wide transcriptional analysis of the mitotic cell cycle. *Molecular Cell*, 2:5-73, 1998.
4. S. Chung, J. Jun, and D. McLeod. Mining gene expression datasets using density-based clustering. In *Proceedings of ACM CIKM International Conference on Information and Knowledge Management*, 2004.
5. R.O. Duda, P.E. Hart, and D.G. Stork. *Pattern Classification (2nd Ed.)*. Wiley, New York, 2001.
6. A. Gasch, and M. Eisen. Exploring the conditional coregulation of yeast gene expression through fuzzy k-means clustering. *Genome Biology*, 3(11):1-22, 2002.
7. T.R. Golub *et al.* Molecular classification of cancer: class discovery and class prediction by gene expression monitoring. *Science*, 286(15):531-537, 1999.
8. D. Jiang, C. Tang, and A. Zhang. Cluster analysis for gene expression data: a survey. In *IEEE Transactions on Knowledge and Data Engineering*, 16(11):1370-1386, 2004.

9. L. Parsons, E. Haque, and H. Liu. Subspace clustering for high dimensional data: a review. *ACM SIGKDD Explorations Newsletter*, 6(1):90-105, 2004.
10. G. Salton and M.J. McGill. *Introduction to modern information retrieval*. McGraw-Hill, 1983.
11. P. Tamayo *et al.* Interpreting patterns of gene expression with self organizing maps. In *Proceedings of National Academy of Science*, 96(6):2907-2912, 1999.
12. C. Tang, A. Zhang, and J. Pei. Mining phenotypes and informative genes from gene expression data. In *Proceedings of the 9th ACM SIGKDD International Conference on Knowledge Discovery and Data Mining*, 2003.
13. H. Wang, W. Wang, J. Yang, and P. S. Yu. Clustering by pattern similarity in large data sets. In *Proceedings of ACM SIGMOD International Conference on Management of Data*, 2002.
14. Y. Xu, V. Olman, and D. Xu. Clustering gene expression data using a graph-theoretic approach: an application of minimum spanning trees. *Bioinformatics*, 18(4):536-545, 2002.
15. M.J. Zaki, and M. Peters. CLICKS: Mining subspace clusters in categorical data via K-partite maximal cliques. In *Proceedings of International Conference on Data Engineering*, 2005.

Bayesian Hierarchical Models for Serial Analysis of Gene Expression

Seungyoon Nam[1,2], Seungmook Lee[3], Sanghyuk Lee[2], Seokmin Shin[1],
and Taesung Park[3,*]

[1] Interdisciplinary Program in Bioinformatics, Seoul National University,
Seoul, 151-742, Korea
[2] Division of Molecular Life Sciences, Ewha Womans University, Seoul 120-750, Korea
[3] Department of Statistics, Seoul National University, Seoul, 151-742, Korea
{seungyoon.nam,lee.seungmook}@gmail.com,
sanghyuk@ewha.ac.kr, {sshin,tspark}@snu.ac.kr

Abstract. In the Serial Analysis of Gene Expression (SAGE) analysis, the statistical procedures have been performed after aggregation of observations from the various libraries for the same class. Most studies have not accounted for the within-class variability. The identification of the differentially expressed genes based on the class separation has not been easy because of heteroscedasticity of libraries. We propose a hierarchical Bayesian model that accounts for the within-class variability. The differential expression is measured by a distribution-free silhouette width which was first introduced into the SAGE differential expression analysis. It is shown that the silhouette width is more appropriate and is easier to compute than the error rate.

Keywords: SAGE, Serial Analysis of Gene expression, Bayesian hierarchical model.

1 Introduction

Gene expression patterns show the complex aspects according to temporal and spatial factors such as cell-specific gene expression, tissue-specific expression, physiological stimuli induced gene expression and etc. The various gene expression patterns lead to the gene product quantities. In other words, if the condition around a cell is changed by infection, the cell leads to the change of its shape and behavior that results from variation of quantity of gene product. One of the main purposes of the gene expression study is to find a gene expression quantity induced by its surroundings.

Many studies to quantify or measure gene expression patterns have been performed. For example, the microarray [1] and the Serial Analysis of Gene Expression (SAGE) [2], [3] have been widely used to interrogate the population of

---

* To whom correspondence should be addressed.

M.M. Dalkilic, S. Kim, and J. Yang (Eds.): VDMB 2006, LNBI 4316, pp. 29–39, 2006.
© Springer-Verlag Berlin Heidelberg 2006

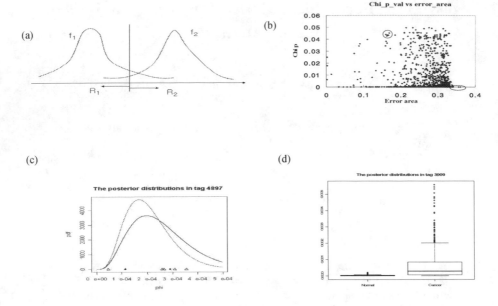

Fig. 1. (a) Error area (see the methods section). The two distributions are f1 and f2. Their corresponding correct classified regions are R1 and R2 respectively. (b) The p-values of the chi-square tests and the error areas for the 1056 tags which are the WinBUGS reported tags among the ones satisfying chi square p value less than 0.05 were plotted. There are two ovals: right lower one and left upper one (see the text). (c) An example from the tags in the left upper oval in the (b). The solid line is the cancer class pooled posterior distribution, and the dashed line is the normal class pooled posterior distribution. The triangles are individual data points for the cancer and the solid circles ones for the normal class. (d) An example from the tags in the right lower oval in (b). The boxplots are plotted against 500 samples from each class posterior distribution.

mRNAs transcribed during biological process in the cells. The data from these techniques can suggest significant insights into the biology of the cell.

Microarray is used to monitor qualitative expression levels of thousands of genes simultaneously. On the other hand, SAGE is used to quantitatively analyze the relative differential abundances of thousands of gene transcription (mRNAs) from a cell or tissue sample. The differences between microarray and SAGE are as follows. First, SAGE uses sequencing as opposed to competitive hybridization. Second, while the expression level of microarray is a measure of fluourescence and is loosely continuous, SAGE has data in the form of counts on gene expression, potentially allowing for a different type of "quantitative" comparison. Third, SAGE is an "open" technology in that it can provide information about all of the genes in the sample. Microarrays, by contrast, are "closed" in that we only get information about the genes that have been printed on the array [4].

In the SAGE analysis, the statistical procedures have been performed after aggregation of observations from the various libraries for the same class, creating a pseudo-library [5], [6]. Most studies have not counted on the within-class variability such as biological variability among individuals within a class, for example, different patients having the same cancer diagnosis. The traditional statistical analysis such as the chi-square test has been performed under the assumption that the expression data from the same class are homogeneous, and as a result the within-class variability has not been taken into consideration.

Recently, Vencio *et al.* (2004) considered the within-class variability under Bayesian frame and proposed a measure of separation called error rate between two class distributions. They accounted for the within-class variability by applying the Beta-Binomial model for expression proportion per tag. However, their approach required numerical computations and set up a boundary condition amenable to relieve the computational intensity of calculating the posteriori density of a class. The error rate, a measure of separation, also needs numerical integrations. The arbitary boundary condition and numerical integration are prone to the serious error.

In order to overcome the burden of numerical integrations and the problem of selecting a boundary condition of Vencio *et al.* (2004), we propose a Bayesian hierarchical model that accounts for the within-class variability but requires a simple Monte Carlo sampling strategy. Our class posterior distribution of expression proportion is approximated by proportion samples of each tag over libraries. The proposed method avoids an intensive numerical integration and does not make an arbitrary adjustment for differential expressions.

In addition, as a class separation measure for two-class classification, we propose using an average silhouette width related to the number of poorly classified observations from two class distributions. The silhouette width is more appropriate and is easier to compute than the error rate.

In Section 2, the proposed Bayesian hierarchical model that accounts for the within-class variability is described. Section 3 illustrates the proposed model using the data from the three normal breast libraries and the five breast cancer libraries from the CGAP (The Cancer Genome Anatomy Project, http://cgap.nci.nih.gov). Discussions and final conclusions are given in Section 4.

2 Materials and Methods

2.1 Bayesian Hierarchical Model

We assume that there are two classes for comparison that are denoted by k (=1, 2) . The CGAP data with two classes is summarized in Table 1: normal class with three libraries and breast cancer class with five libraries. For a given class, suppose that there are J libraries and that the jth library has n_j tags. Let θ_j be the tag proportion in the jth library and y_j be its corresponding tag frequency. Each tag data from libraries j is assumed to have the following independent binomial distribution:

$$y_j \sim Bin(n_j, \theta_j) \tag{1}$$

where the total number of the tag frequency n_j is assumed to be known. The parameters θ_j are assumed to follow the independent beta distribution:

$$\theta_j \sim Beta(\alpha, \beta) \tag{2}$$

We assign a noninformative hyperprior distribution to reflect our ignorance about the unknown hyperparamters, which means "let the data speak by themselves". We assign a proper prior distribution for the hyperparameters α and β. We also assume $p(\alpha,\beta) \approx p(\alpha)p(\beta)$ approximately, and

$$\alpha \sim \Gamma(0.001, 0.001) \text{ and } \beta \sim \Gamma(0.001, 0.001) \tag{3}$$

Table 1. Data description

Class	Library name in the CGAP	Tags
Normal	SAGE_Breast_normal_organoid_B	58181
	SAGE_Breast_normal_myoepithelium_AP_IDC7	69006
	SAGE_Breast_normal_stroma_AP_1	79152
Cancer	SAGE_Breast_metastatic_carcinoma_B_95-260	45087
	SAGE_Breast_carcinoma_epithelium_AP_IDC7	73410
	SAGE_Breast_carcinoma_B_CT15	21840
	SAGE_Breast_carcinoma_associated_stroma_B_IDC7	68024
	SAGE_Breast_carcinoma_B_BWHT18	50701

2.2 Joint, Conditional, and Marginal Posterior Distributions

The joint posterior distribution of θ_js and other parameters for a tag in the kth class is given as follows [7]:

$$p(\theta, \alpha, \beta \mid y) \propto p(\alpha, \beta)p(\theta \mid \alpha, \beta)p(y \mid \theta, \alpha, \beta)$$

$$\propto p(\alpha, \beta)\prod_{j=1}^{J} \frac{\Gamma(\alpha+\beta)}{\Gamma(\alpha)\Gamma(\beta)}\theta_j^{\alpha-1}(1-\theta_j)^{\beta-1}\prod_{j=1}^{J}\theta_j^{y_j}(1-\theta_j)^{n_j-y_j} \tag{4}$$

$$\text{where } \theta=(\theta_1,..,\theta_j,...,\theta_J) \text{ and } y = (y_1,..,y_j,...,y_J).$$

For simplicity, we omit the subscript k, when the meaning is clear. Given (α,β), the components of θ have independent posterior densities that are the form of $\theta_j^A(1-\theta_j)^B$, that is beta densities, and the joint density is

$$p(\theta \mid \alpha, \beta, y)=$$
$$\prod_{j=1}^{J}\frac{\Gamma(\alpha+\beta+n_j)}{\Gamma(\alpha+y_j)\Gamma(\beta+n_j-y_j)}\theta_j^{\alpha+y_j-1}(1-\theta_j)^{\beta+n_j-y_j-1} \tag{5}$$

It is self-evident that for each $j=1,..,J$, θ_j's conditional posterior distribution is

$$\theta_j \mid \alpha, \beta, \mathbf{y} \sim Beta(\alpha + y_j, \beta + n_j - y_j)\, where\ j = 1, .., J. \tag{6}$$

2.3 Pooled Library Model

We assume two pooled libraries are constructed for each class. Let θ be a tag proportion, y be the total frequency for the tag and n be the total tag count for the pooled library in the kth class. From equation (6),

$$\theta \mid \alpha, \beta, y \sim Beta(a, b),$$
$$where\ a = \alpha + y \quad and \quad b = \beta + n - y. \tag{7}$$

For a tag in library j, we use equation (6) to sample θ_j s by the WinBUGS whose initial values are 0.1 for θ_j, α, and β in equation (6). The samples are aggregated to construct the pooled class library model through the multiple libraries. To estimate the parameters a and b of equation (7) using the aggregated samples, the following equations can be used to get the method of moment estimators [7].

$$a + b = \frac{E(\theta)(1 - E(\theta))}{var(\theta)} - 1 \tag{8}$$
$$a = (a + b)E(\theta), \quad b = (a + b)(1 - E(\theta)).$$

This method of moments can be more easily applied for estimating a class posteriori density than the computer-intensive method for estimating the mode of the a posteriori distribution [8].

2.4 Error Area

The error area is a special case of the expected cost minimization problem with misclassification costs and priors being all 1s. It is also similar to Vencio et al's measure of separation that is defined to be the overlapped region area between two distributions. However, it suffers from the limitation of integration range due to the serious numerical errors.

Let f_1 and f_2 represent the two class distributions and R_1 and R_2 be their corresponding correct regions, respectively. Then, the error area is defined by

$$Error\ area = \int_{R_2} f_1(x)dx + \int_{R_1} f_2(x)dx \tag{9}$$

The error boundary between R_1 and R_2 is defined arbitrarily as the mean of the two distribution means. R_1 is the region between the mean of f_1 and the error boundary. R_2 is the area between the error boundary and the mean of f_2. The schematic graph is shown in Fig. 1 (a).

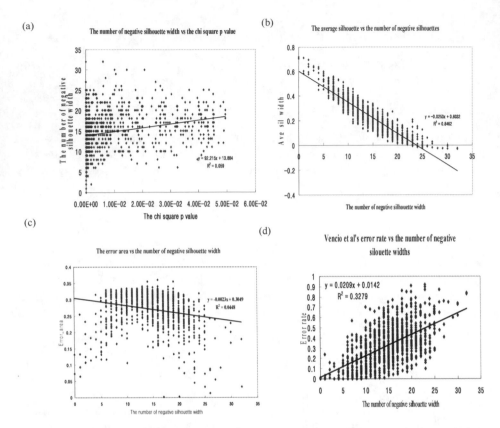

Fig.2. (a) The plot between the number of the negative silhouette width and the p-values of the chi-square test from the 1056 tags. (b) The plot between the average silhouette width and the number of negative silhouette width from the 1056 tags. (c) The error area versus the number of negative silhouette width. (d) The error rate of Vencio *et al* versus the corresponding number of negative silhouette width is plotted against the 1056 tags.

This measure indicates how well the two pooled distributions are separated between the normal and cancer classes for each tag. The more separable between the two distributions, the more differentiable the expressions are.

2.5 Silhouette Width

Silhouette width is used as a measure for checking the adequacy of a clustering analysis [9]. We propose using the silhouette width as a measure of separation between the two pooled class posterior distributions. Suppose that the two data sets are generated from the two posterior distributions. For the ith observation its silhouette width is defined as

$$silhouette\ width: sil_i = (b_i - a_i) / \max(a_i, b_i), \qquad (10)$$

where a_i denotes the average distance between observation i and all other observations in the same class to which observation i belongs, and b_i denotes the average distance between the observation i and all observations belonging to the other class. Intuitively, the tag with a large silhouette width sil_i is well-separated; those with small sil_i tend to lie between the two distributions. A tag with a negative silhouette width lies in the overlapped region between the two distributions.

The average silhouette width from the two class posterior distributions can be used as a summary measure of separation. The more overlapped between the two classes, the more negative average silhouette width. On the other hand, the more separated between the two classes, the more positive average silhouette width. A negative silhouette width is mainly resulted from poor separation between the two pooled distributions because of their large variances and a small location parameter shift. The number of the negative silhouette widths corresponds to the sum of the numbers of the false positives and false negatives. Since the silhouette width is free to distribution, the number of the negative silhouette width can be an alternative measure for separation.

3 Results

3.1 Overview

We illustrate the proposed hierarchical Bayesian model using the three normal breast libraries and the five breast cancer libraries from the CGAP (http://cgap.nci.nih.gov/). Each library is assumed to be independent from each other. We first performed the traditional chi-square tests for the unique 21072 tags. The 5% significance level resulted in 2332 significant tags that are differentially expressed between cancer and normal cells.

We applied the Bayesian hierarchical model to the screened 2332 tags. The model described in the methods section was applied to each tag. The beta-binomial model was constructed and its hyperparameters were assumed to follow a noninformative gamma distribution. The posterior distribution of the tag proportion θ_j was derived by equation (6) and θ_j s were sampled by the WinBUGS. Collecting all $\theta = (\theta_1, ..., \theta_j, ..., \theta_J)$ s in the same class, we estimated the class pooled library distribution using these samples. All computations were performed using the WinBUGS (http://www.mrc-bsu.cam.ac.uk/bugs/winbugs/contents.shtml) and R2WinBUGS [10] (http://www.stat.columbia.edu/~gelman/bugsR/) developed under the R environment (http://cran.r-project.org).

From the 2332 tags the Bayesian method using WinBUGS yielded the 1056 pairs of the posterior distributions for the normal and cancer classes. The remaining tags were unreported due to WinBUGS's program errors caused by the sparseness of data. The 1056 unique tags were studied to examine their significance with regard to their normal and cancer pooled distributions.

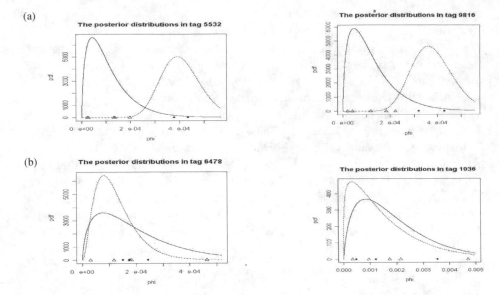

Fig. 3. (a) Examples with a high average silhouette width. Tag 5532 has the posterior distributions with the chi-square p value 1.50E-10 and with an average silhouette width 0.718602, while tag 9816 has 3.49E-06 and 0.5826, respectively. (b) Examples with a low average silhouette width. Tag 6478 has the posterior distributions with the chi-square p value 0.0145 and an average silhouette width -0.02648, while tag 1036 has 0.00696 and -0.02493, respectively. The solid line is a cancer posterior distribution and the dashed line is a normal posterior distribution. The solid circles are individual data points for the normal class and the triangles are ones for the cancer class.

3.2 Error Area

For each tag, the error area was computed from the two class posterior distributions. The error areas of the 1056 unique tags were obtained. We investigated whether or not the error area would be a good separation measure between the two class distributions. In Figure. 1 (b), we did not find a strong linear relationship between the p-value of the chi-square test and the error area. There are two ovals in Figure 1 (b): the left upper one and the right lower one. The tags in the left upper one are expected to be well separated because of their low error areas. However, the posterior distributions for tag 4897 in the left upper oval show that two classes are not well separated, as shown in Figure 1. (c). Also, the tags in the right lower oval are expected to be poorly separated because of their large error areas. However, the posterior distributions for tag 3909 in this oval show that two classes are well separated, as shown in Figure 1. (d). Thus, the error area measure did not seem to represent how well two classes are separated. Further, the error area did not show any relationship with the p-value of the chi-square test assuming a homogeneous pooled class library.

3.3 Silhouette Width

The silhouette width in equation (10) considers both within-group and between-group variances. For each normal and tumor class, we generated 25 samples from the pooled posterior distributions for the 1056 tags. Thus, there are 50 samples for each tag. The number of negative silhouette width and the average silhouette width were calculated for each tag.

We inspected the relationship between the p-value of the chi-square test and the number of negative silhouette widths for all tags, and also the relationship between the number of negative silhouette width and the average silhouette. In Figure 2, the average silhouette width and the number of negative silhouette width show a strong linear relationship, while there is no linear relationship between the number of the negative silhouette width and the p-value of the chi-square test. There is no strong linear relationship between the error area and the number of negative silhouette width. In Figure 2(d), the error rate of Vencio *et al* has a weak linear relationship with the number of negative silhouette width. Thus, the traditional chi-square test to find

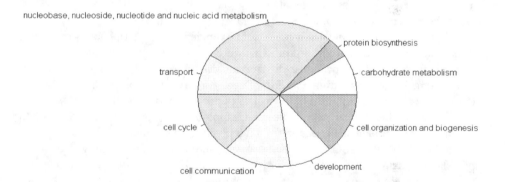

GO Biological process

Fig. 4. The functional enrichment analysis was performed against the biological process

significant expressed genes might have different results from those based on the silhouette width. We think the silhouette width can be a good choice to represents the two distributions' separation. Figure 3 illustrates several examples to show good and poor separation of the two distributions based on the silhouette width.

3.4 Functional Enrichment Analysis

The 1056 tags were sorted in descending order of their total average silhouette width and the upper 25 percentiles (264 tags) were selected. From these tags, the 114 tags were identified by the 2 fold ratio of the cancer class mean to the normal class mean.

The 114 tags were assigned to the 113 unique HUGO genes and the 72 unique GO terms through the ECgene cluster [11]. The cell cycle GO terms plus the cell communication GO terms are one of the dominant GO terms in Figure 4. This result is reasonable in the sense that the progress of tumor-cell invasion and metastasis are understood as cell migration and cancer cell migration is regulated by integrins, matrix-degrading enzymes, cell-cell adhesion molecules and cell-cell communication [12].

4 Discussion

We proposed a Bayesian hierarchical model accounting for within-class variability. The proposed approach avoids heavy numerical integrations of the previous methods by using Monte Carlo sampling strategy. Although it may require more computing times, the proposed approach is much easier to apply.

The silhouette width method, which has been used in the clustering quality analysis, was first, as far as we know, introduced into the SAGE analysis. The differential expression study through multiple libraries suffers from the hetero-scedasticity problem. The silhouette width is a distribution-free measure and does not require an intensive computational integration. Note that the silhouette width is based on the distance and uses the numbers of the false positives and false negatives.

The proposed Bayesian model can be easily extended to the multiple-class problems. The silhouette width can also be used as a measure of separation in this extension.

In many practical applications, differentially expressed genes in the statistical sense might be different from the differentially expressed genes in the biological sense. The posterior distributions of two classes provide more useful information than the simple p-values of the tests. Thus, our approach can provide more insight to investigate this problematic situation from the posterior distributions.

Unexpectedly, WinBUGS yielded too many unreported tags possibly due to the sparseness of data, which requires a further future study. In addition, we will perform further systematic comparisons for the other methods for SAGE.

Acknowledgements. The work of SL was supported by the Ministry of Science and Technology of Korea through the Bioinformatics Research Program (Grant No. 2005-00201) and the Korea Science and Engineering Foundation through the Center for Cell Signaling Research at Ewha Womans University. The work of TP was supported by the National Research Laboratory (Grant No. M10500000126).

References

1. Schena, M., Shalon, D., Davis, R.W., Brown, P.O.: Quantitative monitoring of gene expression patterns with a complementary DNA microarray. Science, (1995), 270:467-470
2. Velculescu, V.E., Zhang, L., Vogelstein, B., Kinzler, K.W.: Serial analysis of gene expression. Science, (1995), 270:484-487

3. Bertelsen, A.H., Velculescu, V.E.: High-throughput gene expression analysis using SAGE. Drug Discovery Today, (1998), 3:152-159
4. Baggerly, K.A., Deng, L., Morris, J.S., Aldaz, C.M.: Overdispersed logistic regression for SAGE: modelling multiple groups and covariates. BMC Bioinformatics, (2004), 5:144
5. Ruijter, J.M., Van Kampen, A.H., Baas, F.: Statistical evaluation of SAGE libraries: consequences for experimental design. Physiol Genomics, (2002), 11:37-44
6. Man, M.Z., Wang, X., Wang, Y.: POWER_SAGE: comparing statistical tests for SAGE experiments. Bioinformatics, (2000), 16:953-959
7. Gelman, A., Carlin, J.B., Stern, H.S., Rubin, D.B.: Bayesian Data Analysis. 1st edn. Chapman & Hall/CRC, (1995)
8. Vencio, R.Z., Brentani, H., Patrao, D.F., Pereira, C.A.: Bayesian model accounting for within-class biological variability in Serial Analysis of Gene Expression (SAGE). BMC Bioinformatics, (2004), 5:119
9. Kaufman, S., Rousseeuw, P.J.: Finding Groups in Data: An Introduction to Cluster Analysis. Wiley, New York (1990)
10. Sturtz., S., Ligges., U., Gelman., A.: R2WinBUGS: A Package for Running WinBUGS from R. Journal of Statistical Software, (2005), 12
11. Kim, N., Shin, S., Lee, S.: ECgene: genome-based EST clustering and gene modeling for alternative splicing. Genome Res, (2005), 15:566-576
12. Friedl, P., Wolf, K.: Tumor-cell invasion and migration: Diversity and escape mechanisms. Nature Rev. Cancer, (2003), 3:362-374

Applying Gaussian Distribution-Dependent Criteria to Decision Trees for High-Dimensional Microarray Data

Raymond Wan, Ichigaku Takigawa, and Hiroshi Mamitsuka

Bioinformatics Center, Institute for Chemical Research, Kyoto University, Gokasho, Uji, 611-0011, Japan
{rwan,takigawa,mami}@kuicr.kyoto-u.ac.jp

Abstract. Biological data presents unique problems for data analysis due to its high dimensions. Microarray data is one example of such data which has received much attention in recent years. Machine learning algorithms such as support vector machines (SVM) are ideal for microarray data due to its high classification accuracies. However, sometimes the information being sought is a list of genes which best separates the classes, and not a classification rate.

Decision trees are one alternative which do not perform as well as SVMs, but their output is easily understood by non-specialists. A major obstacle with applying current decision tree implementations for high-dimensional data sets is their tendency to assign the same scores for multiple attributes. In this paper, we propose two distribution-dependant criteria for decision trees to improve their usefulness for microarray classification.

1   Introduction

Biological data presents unique problems due to the high dimensions of the data sets. Experiments are generally costly and in most cases, obtaining samples require ethical approval. Thus, the number of samples is usually much smaller than the number of values collected per sample.

Microarrays are costly, yet high-throughput tools which allow biologists to measure the expression levels of thousands of genes simultaneously from a particular sample. While the number of samples in a typical study may be limited, the dimensions can be as high as several thousand genes.

Both supervised and unsupervised learning methods have been used for microarray analysis in the form of support vector machines (SVM) and hierarchical clustering. On the other hand, decision trees are time-tested machine learning tools that do not classify as well as SVM, but have the advantage of having its results being easily interpreted by someone with limited knowledge of the underlying algorithm. The most important features (genes) in the data set which best separate the data into its respective classes can be returned to the user.

The high-dimensionality of microarray data sets makes this type of data mining difficult using current decision tree implementations. If the number of samples is small, some genes may be identified as being important for classification when in fact they are not. In this paper, we combine the criterion found in one decision tree implementation

M.M. Dalkilic, S. Kim, and J. Yang (Eds.): VDMB 2006, LNBI 4316, pp. 40–49, 2006.

(C4.5 Release 8 [1]) with two distribution-dependent splitting criteria which assume a Gaussian distribution. While these methods on their own are not novel, we combine them to improve the statistical reliability of decision trees for microarray data.

This paper is organized as follows. Section 2 covers some relevant background, including certain aspects about C4.5. Section 3 gives a list of related work in decision trees and microarray classification. Our method is detailed in Section 4, followed by experimental results with both artificial and real data in Section 5. Finally, Section 6 closes this paper by summarizing our method.

2 Background

A data set $\mathcal{D}$ consists of m examples with n attributes (or features) each. The examples are enumerated from S_1 to S_m while the attributes range from A_1 to A_n. The value v_{ij} corresponds to the value at example S_i and attribute A_j. Each example also has an associated class from the feature class set C_i. We focus on binary classification, so there are only two classes, $C_i = \{c_1, c_2\}$.

A decision tree is built through the recursive partitioning of $\mathcal{D}$ into smaller subsets $\mathcal{D}_1, ..., \mathcal{D}_k$. Fundamental to decision tree construction is a splitting criterion. The splitting criterion is called upon whenever a new node is added, effectively splitting the data set further. In the case of continuous values, the splitting criterion also provides the value v to split on (the cut point). Most implementations create two children, so that one branch has the samples which satisfy $A_k \leq v$; the remaining ones satisfy $A_k > v$.

The criterion employed by C4.5 is the gain ratio, an extension to information gain [2]. Based on information theory, the gain ratio selects the attribute which creates subsets with the most uniformity in class. Release 8 of C4.5 applies the Minimum Description Length (MDL) principle to reduce bias towards selecting continuous attributes [1].

As the number of attributes increases compared to the number of samples, the chance that two attributes will yield the same gain ratio increases. Unless additional rules are in place, the attributes which tie with the highest gain ratio have to be decided arbitrarily. As an example, C4.5 selects the first attribute encountered when processing $\mathcal{D}$ from left-to-right. Further details about decision trees can be found elsewhere [2,3].

3 Related Work

Our method combines two Gaussian distribution-dependant splitting criteria with the gain ratio of C4.5. The aim is to reduce the number of cases where the splitting criteria ties with multiple attributes during tree construction.

Coupling normal distributions with decision tree construction has been considered by others [4]. In their work, the means and standard deviations were compared directly in a pair-wise fashion of all examples. More recent work include DB-CART and QUEST. DB-CART used the joint probability distribution of all attributes with the class label to build a probability density estimate for each class [5]. QUEST (and the algorithm which it builds upon) used an analysis of variance (ANOVA) F-statistic for attribute selection [6]. In contrast to these earlier methods, the focus of our work is high-dimensional biological data with tying attributes, with a particular emphasis on microarray data.

The assumption that genes follow a Gaussian distribution was chosen because it is the most prevalent in the microarray analysis literature. When there are only two classes (as we are also assuming), previous work has applied Student's t-test to validate the results from a clustering algorithm [7] or used it to act as a gene pre-filter [8]. However, opinions on the normality of microarray data are varied. It has been suggested that t-test is favored when dealing with the logarithms of the measurement values [9]. Others have stated that the assumption of normality is safe for untransformed Affymetrix data sets [10]. Alternatives to t-test, which we do not consider in this paper, include less stringent distribution-free (or non-parametric) tests.

With respect to decision trees and microarrays, recursive partitioning (another name for decision trees) has been used on microarray data [11]. The splitting criterion chosen was based on the probability that a tissue was normal and the choices were further refined using cross-validation. In their more recent work, a forest of decision trees was built and the top two levels of multiple trees formed a fingerprint for a particular cancer type [12]. The aim of our method is not to replace methods such as decision forests, but instead to complement them.

RankGene is a software application for gene selection which includes information gain and t-test [13]. Unlike RankGene, we integrate other feature selection methods together, instead of using each one independently.

4   Method

The values in each attribute are assumed to follow two Gaussian distributions representing the two classes. The two normal distributions $\mathcal{N}_1(\mu_1, \sigma_1)$ and $\mathcal{N}_2(\mu_2, \sigma_2)$ apply to the classes c_1 and c_2, respectively, for some attribute A_k. These two distributions are assumed to be independent between genes. We conservatively assume that both of their means and variances are unequal.

The two distribution-dependent splitting criteria that we augment with the gain ratio are Student's t-test and the Kullback-Leibler divergence. An aggregated score, $score(A_k)$, is assigned to each attribute A_k, for selecting the splitting attribute.

This section outlines the pre-selection process of attributes, followed by the procedure to create the two scores, $ttest_s$ and KL_s. The gain ratio of C4.5 is untouched and is referred to as $gain_s$. We then show how these three metrics are combined through ranking to derive $score(A_k)$, for attribute A_k.

4.1   Pre-selecting Attributes

We first omit attributes which do not offer a gain ratio which outweighs its MDL cost. Thus, the gain ratio serves two purposes: calculating $gain_s$ and acting as a Boolean test to exclude attributes from being a splitting attribute.

4.2   Distribution-Dependent Splitting Criteria

The Student's t-test for testing the difference in means exists in various forms. Since we are only concerned with whether or not the two distributions differ, we elected to use a two-tailed, two sample t-test for unequal variance. The t-test score, $ttest_s$, for the current attribute is the absolute value of the t-statistic. Our implementation makes use of routines provided by Numerical Recipes in C [14]. The Kullback-Leibler divergence (or KL divergence) calculates the distance between two probability distributions, usually between a "true" distribution and a theoretical one [15]. Unlike t-test which emphasizes the difference in means, KL divergence places importance in both their means and their variances. Let $p(x)$ and $q(x)$ represent the two probability density functions (pdfs). Assuming a normal distribution, the KL divergence $D(p\|q)$ is given as:

$$D(p\|q) = \int_{-\infty}^{+\infty} p(x) \log \frac{p(x)}{q(x)} dx . \tag{1}$$

If we assume that these two pdfs are normally distributed for the classes c_1 and c_2, then their KL divergence can be simplified to:

$$D(p\|q) = \frac{1}{2\sigma_2^2} (\mu_1 - \mu_2)^2 + \frac{(\sigma_1 - \sigma_2)(\sigma_1 + \sigma_2)}{2\sigma_2^2} + \log\left(\frac{\sigma_2}{\sigma_1}\right) . \tag{2}$$

Since the Kullback-Leibler divergence is non-symmetric, the Kullback-Leibler divergence score KL_s is calculated as half of $D(p\|q) + D(q\|p)$ [16].

4.3 Combining the Scores

The scores from all three metrics are combined to create a unified score which is then used to select the best splitting attribute. We rank each set of scores across all attributes to derive an aggregated score.

The ranks, $gain_r$, $ttest_r$, and KL_r, are assigned from n down to 1. If multiple scores tie, they are all assigned the same rank. Then the next highest score is assigned a rank which skips the tied ones. After ranking each attribute according to its gain ratio, t-statistic, and KL divergence, we introduce three parameters ($\{\alpha_1, \alpha_2, \alpha_3\}$) which weight the sum of the ranks:

$$\text{score}(A_k) = \alpha_1 \times gain_r + \alpha_2 \times ttest_r + \alpha_3 \times KL_r , \text{ such that } \sum_{i=1}^{3} \alpha_i = 1. \tag{3}$$

4.4 Example

We complete this section with a brief example. In Table 1, a data set is shown made up of two classes, A and B. There are five examples with two attributes each. It is obvious that not only is either attribute able to separate the data set into two uniform classes, but x_2 is the better choice. A decision tree algorithm must choose between these two attributes somehow. Obviously, this problem becomes more pronounced when the dimensions of the data set resembles a microarray data set.

Table 1. Example data set with five examples and two attributes based on two classes, A and B

x_1	x_2	C
S_1 1	1	A
S_2 2	2	A
S_3 3	3	A
S_4 4	10	B
S_5 5	11	B

Table 2(a) calculates the gain ratio, t-statistic, and KL divergence for the data in Table 1. Since the gain ratio focuses on the class distribution, it ties at 0.794. However, the other two tests do not tie and x_2 is clearly a better discriminate between the two classes – both the t-statistic and the KL divergence are larger. The ranks are then assigned such that they are the same only for $gain_r$.

The final scores for the attributes are given in Table 2(b) using two sets of parameters. For the first parameter, $\alpha_1 = 1$, so both $ttest_r$ and KL_r will have no effect on the score. Since a tie results, a decision tree implementation must arbitrarily choose one of them to split on. On the other hand, if equal weight is given to all 3 attributes, then x_2 will attain the higher score and be picked as the splitting attribute.

Table 2. Calculation of the scores for x_1 and x_2 using the sample data set of Table 1

A_k	$gain_s$	$gain_r$	$ttest_s$	$ttest_r$	KL_s	KL_r	$\{\alpha_1, \alpha_2, \alpha_3\}$	$score(x_1)$	$score(x_2)$
x_1	0.794	2	3.273	1	16.75	1	$1, 0, 0$	2.00	2.00
x_2	0.794	2	11.129	2	181.75	2	$\frac{1}{3}, \frac{1}{3}, \frac{1}{3}$	1.33	2.00

(a) Calculation of the three measures.   (b) Combining the measures using two different sets of parameters.

5   Results

Experiments were conducted using C4.5 Release 8 and its default parameters (including pruning) on an artificial data set and seven publicly-available microarray data sets.

The seven microarray data sets obtained are shown in Table 3. In our experiments, we first apply inverse cross-validation such that the training set is smaller than the test set. With a training set no larger than 20 samples, we are able to emulate the average size of most microarray studies while still reserving a large portion of the data for testing. Though not the focus of our work, for the sake of completeness, we also demonstrate our method with (normal) cross-validation.

5.1   Artificial Data

Simulations with artificial data demonstrate the ideal scenario for our method. We created data sets made of two types of attributes, with each attribute's values generated independently of other attributes. The first type of attribute contained two overlapping distributions with identical standard deviations, but different means: $\mathcal{N}_1(0, 1)$ and $\mathcal{N}_2(1.5, 1)$. There were 25 to 10,000 attributes of this type. The second type of attribute were distant in mean: $\mathcal{N}_1(0, 1)$ and $\mathcal{N}_2(100, 1)$. The single attribute with distant means was placed at the end of the data set. Recall that when a tie in the best gain ratio occurs, C4.5 chooses the first attribute it encounters with that gain ratio.

Partial results from these experiments, averaged across 50 iterations, are presented in Figure 1. There are three variables to consider: the number of samples for training, the number of samples for testing, and the total number of attributes. Throughout our

Table 3. The microarray data sets used are referred to by their name in this paper. Other information listed include the classes, the number of samples, and the number of genes.

Name	Classes		Number of samples		Number of genes	Citation
	Class 1	Class 2	Proportion	Total	genes	
Colon	Tumor	Normal	40 : 22	(62)	2,000	[7]
Leukemia	ALL	AML	47 : 25	(72)	7,129	[17]
Lung	ADCA	MPM	150 : 31	(181)	12,533	[18]
CNS	Success	Fail	39 : 21	(60)	7,129	[19]
Multiple	Tumor	Normal	190 : 90	(280)	16,063	[20]
Lymph	DLBCL	FL	58 : 19	(77)	7,129	[21]
Prostate	Tumor	Normal	52 : 50	(102)	12,600	[22]

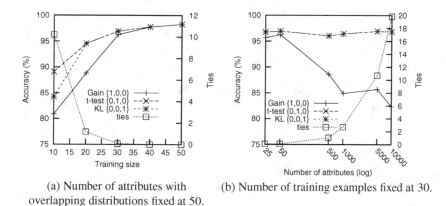

(a) Number of attributes with overlapping distributions fixed at 50.　　(b) Number of training examples fixed at 30.

Fig. 1. Results from experiments with artificial data, repeated 50 times

experiments with artificial data, the size of the test set is held at 1000. In Figure 1(a), we fix the number of attributes with overlapping distributions to 50 and vary the size of the training set from 10 to 50. In Figure 1(b), we fix the training size at 30 examples and vary the number of attributes.

In both graphs, the y-axis on the left indicates the accuracy on the test set and applies to the first three lines of the graph. The first three lines in the legend correspond to the gain ratio ($\{1, 0, 0\}$), t-test ($\{0, 1, 0\}$), and KL divergence ($\{0, 0, 1\}$), respectively. The y-axis on the right indicates the number of attributes whose gain ratios tied and is applicable only to the last line.

Figure 1(a) shows that for a training size of 10 or 20 examples, t-test outperforms the other methods, but its advantage diminishes as the training size continues to increase toward 50. The parameters selected for the normal distributions indicate that our aim is to examine differences in means, not variance. Hence, we expect t-test to be slightly more effective than KL divergence. As the training size increases, the number of tying attributes decreases.

In Figure 1(b), the performance of the gain ratio diminishes as the number of attributes increases. The increase in attributes corresponds to an increase in the number

Table 4. Experiments with inverse cross-validation. Results are absolute accuracies for $\{1, 0, 0\}$ only and relative to this baseline otherwise. All values are averaged across 50 trials.

Training size	$\{\alpha_1, \alpha_2, \alpha_3\}$	Colon	Leukemia	Lung	CNS	Multiple	Lymph	Prostate	Avg
	$\{1, 0, 0\}$	59.5	58.8	75.8	52.6	59.3	65.5	51.6	60.5
	$\{0, 1, 0\}$	+1.4	+10.2	+4.6	+2.2	+2.0	+3.7	+12.1	+5.2
		(1.81)	**(13.20)**	**(9.27)**	**(3.94)**	**(8.79)**	**(5.27)**	**(23.26)**	
	$\{0, 0, 1\}$	+0.8	+6.1	+6.9	+4.9	-0.7	+3.1	+3.4	+3.5
10		(0.89)	**(10.16)**	**(17.50)**	**(6.09)**	(-2.94)	**(4.04)**	**(6.92)**	
	$\{\frac{1}{3}, \frac{1}{3}, \frac{1}{3}\}$	+2.3	+11.4	+5.3	+3.4	+1.5	+6.6	+7.6	+5.4
		(3.23)	**(16.26)**	**(11.94)**	**(5.43)**	**(6.00)**	**(9.15)**	**(16.87)**	
	$\{0, \frac{1}{2}, \frac{1}{2}\}$	+2.0	+11.7	+5.6	+3.6	+1.7	+7.0	+9.4	+5.8
		(2.74)	**(15.09)**	**(13.70)**	**(5.60)**	**(6.76)**	**(9.59)**	**(17.84)**	
	$\{1, 0, 0\}$	68.2	82.7	81.1	55.2	63.3	74.4	68.0	70.4
	$\{0, 1, 0\}$	-2.0	-3.0	+5.4	+0.5	+0.5	+3.8	+5.8	+1.6
		(-2.33)	(-3.11)	**(11.92)**	(0.65)	(1.55)	**(5.28)**	**(7.55)**	
	$\{0, 0, 1\}$	-2.0	-6.0	+3.9	+1.9	-1.8	-2.2	-7.4	-1.9
20		(-1.87)	(-5.70)	**(9.13)**	(2.29)	(-6.24)	(-2.58)	(-8.59)	
	$\{\frac{1}{3}, \frac{1}{3}, \frac{1}{3}\}$	+0.9	+2.5	+6.8	+0.6	+1.0	+4.3	+5.4	+3.1
		(0.93)	(2.51)	**(15.72)**	(0.68)	**(3.79)**	**(5.96)**	**(7.97)**	
	$\{0, \frac{1}{2}, \frac{1}{2}\}$	-0.1	+1.2	+7.1	+0.8	+1.1	+4.3	+5.3	+2.8
		(-0.13)	(1.32)	**(16.88)**	(0.83)	**(3.85)**	**(5.38)**	**(8.04)**	

of tying attributes. The performance of both t-test and KL divergence remain above the 95% mark.

5.2   Inverse Cross-Validation

Results from applying inverse cross-validation to the seven microarray data sets (shown in Table 3) are reported in Table 4. As with the experiments with artificial data, 50 trials were performed. Results are presented from the perspective of the size of the training set. That is, if there are 10 training examples for the Colon data set of 62 examples, then six-fold inverse cross-validation was employed.

Absolute accuracies are shown for the baseline method ($\{1, 0, 0\}$). All remaining accuracies are relative to the baseline, with positive and negative signs indicating gain or loss. In parentheses below each relative accuracy is the t-statistic for the hypothetical test against the baseline over the 50 trials. For 50 trials, the two-tailed t-statistic is 2.6800 for a 99% confidence interval.

The table indicates that using t-test and KL together improves the prediction accuracy over the baseline. When the training data size is 10, the accuracy of using t-test and KL was much better than the baseline, being statistically significant in all 14 (two conditions times seven datasets) cases. This is not true if either t-test or KL are used in isolation.

If we focus on the two sets of parameters which each have at least two parameters being non-zero, note that the number of significant accuracies decreases as the training size increases (i.e., from 14/14 to 8/14). Thus, the modifications to C4.5 worked better for smaller data sets, the size of which are the same as that of the datasets frequently seen in currently available microarray databases.

In additional experiments (not shown), we compared our method with three machine learning algorithms found in WEKA [3]: Naive Bayes (NB), support vector machines (SVM), and decision trees (J4.8). All three were executed using their default values. As J4.8 is essentially a Java implementation of C4.5, no difference in accuracy was found. However, NB and SVM achieved an average improvement across the 7 data sets of 7.0% and 12.0%, respectively, for a training size of 10 examples. For 20 examples, their respective improvements were 2.5% and 10.5%. Thus, for both training sizes, SVM yields better accuracies than our modifications to C4.5.

5.3 Normal Cross-Validation

Even though our main focus is microarray data sets with few samples, we also applied normal cross-validation in order to provide a complete picture of its effectiveness. We apply both five and ten fold cross-validation to all seven data sets and report our findings in Table 5. In this table, the first column indicates the number of folds created. The first obvious difference is the overall improvement in the baseline accuracy due to the increase in training data. Even so, statistically significant improvement in accuracy is still achieved for both fold sizes using our method. Focussing again on the executions where at least two parameters were non-zero, the number of statistically significant improvements were 7/14 and 4/14 for five and ten fold cross-validation, respectively. While accuracy is sometimes reduced, note that in none of the cases is the decrease statistically significant.

As for the WEKA implementations of Naive Bayes and SVM (not shown), Naive Bayes performed worse than the baseline C4.5 (-1.7% and -2.1% for five and ten fold cross-validation) while SVM performed better (10.5% and 10.6%, respectively).

6   Discussion

We have shown how the accuracy of decision trees can be improved if the gain ratio of C4.5 is juxtaposed with two distribution-dependent splitting criteria, Student's t-test and Kullback-Leibler divergence. Improvement was up to 10% compared to the baseline if the training size had about 10 examples. The motivation for this work was to address the issue of ties in the gain ratio during tree induction of high-dimensional data sets such as microarray data. Our modifications to the original implementation does not reduce accuracy, but either improve or leave it unchanged in a statistically significant manner. We also show that our method improves C4.5 slightly when normal cross-validation is employed.

Additional experiments showed that Naive Bayes and SVM outperform both the baseline C4.5 and our method. However, our paper has not shown how the output of decision trees is more meaningful over algorithms such as SVM. In the context of microarray data, decision trees readily outputs the most important genes used for classification, instead of a description of a hyperplane between the two classes. In the future, we would like to evaluate the effectiveness of our method from the perspective of the genes that it identifies with the aid of some domain expertise.

Table 5. Experiments with cross-validation. Results are absolute accuracies for $\{1, 0, 0\}$ only and relative to this baseline otherwise. All values are averaged across 50 trials.

Folds $\{\alpha_1, \alpha_2, \alpha_3\}$	Data Set							Avg
	Colon	Leukemia	Lung	CNS	Multiple	Lymph	Prostate	
$\{1, 0, 0\}$	76.5	84.8	93.4	56.2	78.5	79.2	82.4	78.7
$\{0, 1, 0\}$	-2.5	+2.5	+2.1	+0.8	-1.8	+11.4	+1.9	+2.1
	(-2.42)	**(3.96)**	**(7.78)**	(0.66)	(-3.96)	**(15.43)**	**(2.90)**	
$\{0, 0, 1\}$	-6.6	+2.9	-1.0	+2.5	-3.3	-5.1	-10.3	-3.0
5	(-7.12)	**(4.64)**	(-2.70)	(2.07)	(-7.26)	(-5.91)	(-13.72)	
$\{\frac{1}{3}, \frac{1}{3}, \frac{1}{3}\}$	-3.6	+3.6	+1.9	+4.0	-0.1	+2.5	-0.4	+1.1
	(-4.58)	**(5.95)**	**(8.21)**	**(3.63)**	(-0.13)	**(3.62)**	(-0.73)	
$\{0, \frac{1}{2}, \frac{1}{2}\}$	-3.6	+6.0	+3.6	+2.0	-0.2	+3.4	+1.2	+1.8
	(-4.73)	**(10.41)**	**(14.41)**	(1.99)	(-0.51)	**(5.25)**	(1.95)	
$\{1, 0, 0\}$	77.5	80.4	93.7	58.3	78.7	81.1	83.3	79.0
$\{0, 1, 0\}$	+0.6	+6.4	+2.2	-2.8	-2.0	+12.1	+0.9	+2.5
	(0.68)	**(10.20)**	**(9.44)**	(-2.69)	(-5.14)	**(21.72)**	(1.52)	
$\{0, 0, 1\}$	-5.2	+10.1	-1.2	+3.5	-3.2	-6.0	-11.9	-2.0
10	(-6.72)	**(16.97)**	(-4.03)	**(3.73)**	(-6.94)	(-9.59)	(-18.23)	
$\{\frac{1}{3}, \frac{1}{3}, \frac{1}{3}\}$	-4.9	+7.8	+1.7	+1.5	-0.3	+0.3	-1.7	+0.6
	(-6.38)	**(12.69)**	**(6.81)**	(1.39)	(-0.88)	(0.40)	(-3.52)	
$\{0, \frac{1}{2}, \frac{1}{2}\}$	-2.7	+11.3	+3.7	-0.1	+0.3	+1.5	+0.6	+2.1
	(-3.89)	**(23.32)**	**(16.14)**	(-0.07)	(0.76)	(2.55)	(1.14)	

The aim of our work is to improve the utility of decision trees for high-dimensional biological data sets such as microarrays. Combined with other paradigms such as decision forests, we envision that decision trees can be a viable alternative to other classification algorithms. Unlike algorithms like SVM which only provide a classification rate to the user, decision trees have the ability to indicate which genes separated the classes the best. This is an important feature which will contribute to the data mining of microarray data.

Acknowledgements. This work was supported in part by the Bioinformatics Education Program "Education and Research Organization for Genome Information Science" and Kyoto University's 21st Century COE Program "Knowledge Information Infrastructure for Genome Science" with support from MEXT (Ministry of Education, Culture, Sports, Science and Technology) Japan.

References

1. Quinlan, J.R.: Improved use of continuous attributes in C4.5. Journal of Artificial Intelligence Research **4** (1996) 77–90 Source available from: http://www.rulequest.com/Personal/.
2. Quinlan, J.R.: C4.5: Programs for Machine Learning. Morgan Kaufmann Publishers (1993)
3. Witten, I.H., Frank, E.: Data Mining: Practical machine learning tools and techniques with Java implementations. Second edn. Morgan Kaufmann Publishers (2005)

4. Luo, R.C., Scherp, R.S., Lanzo, M.: Object identification using automated decision tree construction approach for robotics applications. Journal of Robotic Systems **4**(3) (1987) 423–433
5. Shang, N., Breiman, L.: Distribution based trees are more accurate. In: Proc. International Conference on Neural Information Processing. (1996) 133–138
6. Loh, W.Y., Shih, Y.S.: Split selection methods for classification trees. Statistica Sinica **7** (1997) 815–840
7. Alon, U., et al.: Broad patterns of gene expression revealed by clustering analysis of tumor and normal colon tissues probed by oligonucleotide arrays. Proc. National Academy of Sciences USA **96**(12) (1999) 6745–6750 Data: http://microarray.princeton.edu/oncology/affydata/index.html.
8. Yeung, K.Y., Bumgarner, R.E., Raftery, A.E.: Bayesian model averaging: development of an improved multi-class, gene selection and classification tool for microarray data. Bioinformatics **21**(10) (2005) 2394–2402
9. Wit, E., McClure, J.: Statistics for Microarrays. John Wiley & Sons Ltd. (2004)
10. Giles, P.J., Kipling, D.: Normality of oligonucleotide microarray data and implications for parametric statistical analysis. Bioinformatics **19**(17) (2003) 2254–2262
11. Zhang, H., Yu, C.Y., Singer, B., Xiong, M.: Recursive partitioning for tumor classification with gene expression microarray data. Proc. National Academy of Sciences USA **98**(12) (2001) 6730–6735
12. Zhang, H., Yu, C.Y., Singer, B.: Cell and tumor classification using gene expression data: Construction of forests. Proc. National Academy of Sciences USA **100**(7) (2003) 4168–4172
13. Su, Y., Murali, T.M., Pavlovic, V., Schaffer, M., Kasif, S.: RankGene: identification of diagnostic genes based on expression data. Bioinformatics **19**(12) (2003) 1578–1579 Software available from http://genomics10.bu.edu/yangsu/rankgene/.
14. Press, W.H., Teukolsky, S.A., Vetterling, W.T., Flannery, B.P.: Numerical Recipes in C: The Art of Scientific Computing. Second edn. Cambridge University Press (1999)
15. Kullback, S., Leibler, R.A.: On information and sufficiency. Annals of Mathematical Statistics **22**(1) (1951) 79–86
16. Jeffreys, H.: An invariant form for the prior probability in estimation problems. Proc. Royal Society of London (A) **186** (1946) 453–461
17. Golub, T.R., et al.: Molecular classification of cancer: Class discovery and class prediction by gene expression monitoring. Science **286**(5439) (1999) 531–537 Data: http://www.broad.mit.edu/cgi-bin/cancer/datasets.cgi.
18. Gordon, G.J., et al.: Translation of microarray data into clinically relevant cancer diagnostic tests using gene expression ratios in lung cancer and mesothelioma. Cancer Research **62**(17) (2002) 4963–4967 Data: http://www.chestsurg.org/publications/2002-microarray.aspx.
19. Pomeroy, S.L., et al.: Prediction of central nervous system embryonal tumour outcome based on gene expresion. Nature **415**(6870) (2002) 436–442 Data: http://www.broad.mit.edu/cgi-bin/cancer/datasets.cgi.
20. Ramaswamy, S., et al.: Multiclass cancer diagnosis using tumor gene expression signatures. Proc. National Academy of Sciences USA **98**(26) (2001) 15149–15154 Data: http://www.broad.mit.edu/cgi-bin/cancer/datasets.cgi.
21. Shipp, M.A., et al.: Diffuse large B-cell lymphoma outcome prediction by gene-expression profiling and supervised machine learning. Nature Medicine **8**(1) (2002) 68–74 Data: http://www.broad.mit.edu/cgi-bin/cancer/datasets.cgi.
22. Singh, D., et al.: Gene expression correlates of clinical prostate cancer behavior. Cancer Cell **1**(2) (2002) 203–209 Data: http://www.broad.mit.edu/cgi-bin/cancer/datasets.cgi.

A Biological Text Retrieval System Based on Background Knowledge and User Feedback

Meng Hu and Jiong Yang

EECS, Case Western Reserve University, Cleveland, OH
{meng.hu, jiong.yang}@case.edu

Abstract. Efficiently finding the most relevant publications in large corpus is an important research topic in information retrieval. The number of biological literatures grows exponentially in various publication databases. The objective of this paper is to quickly identify useful publications from a large number of biological documents. In this paper, we introduce a new iterative search paradigm that integrates biomedical background knowledge in organizing the results returned by search engines and utilizes user feedbacks in pruning irrelevant documents by document classification. A new term weighting strategy based on Gene Ontology is proposed to represent biomedical literatures. A prototype text retrieval system is built on this iterative search approach. Experimental results on MEDLINE abstracts and different keyword inputs show that the system can filter a large number of irrelevant documents in a reasonable time while keeping most of the useful documents. The results also show that the system is robust against different inputs and parameter settings.

1 Introduction

Searching for relevant publications from large literature corpora is a frequent task to biomedical researchers and biologists. With the abundance of biomedical publications available in digital libraries in recent years, efficient text retrieval becomes a more challenging task. PubMed [1] now contains over 14 million publications. It is crucial to efficiently and accurately identify those documents most relevant to users' interests from such large document collections. One limiting factor of traditional search engine technology is the low precision of the results returned. When users search by a few keywords, a large number of matched results could be returned. Users spend a significant amount of time browsing these results to find documents they are truly interested in. Biological text search engine, e.g. PubMed [1] and eTBLAST [2], also suffers from this problem. In most cases it is impossible for biologists to manually read every returned document, thus losing many truly relevant documents.

Much work has been done to improve the efficiency and effectiveness of literature retrieval in both the public domain and the biomedical discipline. For example, document ranking is introduced for indexing entries in large literature collections. PageRank [3] and HITS [4] are both citation-based scoring functions

M.M. Dalkilic, S. Kim, and J. Yang (Eds.): VDMB 2006, LNBI 4316, pp. 50–64, 2006.
© Springer-Verlag Berlin Heidelberg 2006

for evaluating documents importance. [5] presented a method that ranks documents in MEDLINE using the differences in word content between MEDLINE entries related to a topic and the whole of MEDLINE. On the other hand, text categorization has been studied to organize the search results. In [6], a machine learning model based on text categorization is built to identify high-quality articles in a specific area of internal medicine. SOPHIA [7] is an unsupervised distributional clustering technique for text retrieval in MEDLINE. However, none of the above work considers the domain knowledge of their problem domains in organizing search results.

In this paper, we propose a new iterative searching paradigm that aims to improve the biological literature search by incorporating biological domain knowledge and user feedbacks. First a set of documents returned by a keyword-based search is organized in a clustering manner, then users provide objective evaluations on a small set of representative documents selected from these document clusters. Biological domain knowledge described in a controlled vocabulary is integrated to help the document clustering process. Next the system takes advantage of user feedbacks to refine the document set by classification. Those documents classified as relevant go to the next iteration while the documents that are classified as irrelevant will be pruned out. Users can stop the iterative search at any time if the number of remaining documents is small enough for them to review, or the search process automatically terminates if a pre-defined number of documents is reached. In this system, the number of documents examined by users is significantly reduced and the size of the retrieved document set also shrinks with the help of the pruning process. This approach is particularly useful when there is a large amount of results returned for a keywords-based search. Furthermore, a prototype system is developed based on this approach to illustrate its effectiveness.

Since our text retrieval system focuses on the biological domain, we believe that background knowledge in this area could benefit the document clustering process, and add explanatory power to the organization of documents. The domain knowledge we exploit in this paper is Gene Ontology [8]. Gene Ontology is a structured, controlled vocabulary that describes gene products in terms of their associated biological processes, cellular components, and molecular functions. We consider Gene Ontology as a hierarchical organization of biological concepts, and incorporate this hierarchical structure in measuring the similarity between biological publications. A weighting scheme of GO terms is proposed to represent documents. Document classification is used to refine the document set. To utilize the user feedbacks in this step, the documents that are evaluated as relevant by users are used as the positive examples in training set.

Document clustering and classification has been an active research area in the past decade. [9] proposed an algorithm towards large scale clustering of MEDLINE abstracts based on statistical treatment of terms and stemming. However, the method proposed does not consider biological knowledge in clustering process. In [10], an idea of "theme" was formalized in a set of documents, and an algorithm was given to produce themes and to cluster documents according to these themes. The limitation of this work is that their method was only validated on a specific sub-domain of biological literature. Ontology-based text clustering is one of the works most relevant to this paper. In [11], a core ontology WordNet is integrated in text clustering process as background knowledge. Concepts in the core ontology are compiled into the representations of text documents. However, their methods may not work for specific biomedical domain, and the formal concept analysis used for conceptual clustering is slow and impractical in real applications. Therefore, in this paper, a new measurement based on a biomedical ontology is proposed to assess the similarity between biological publications. In [12], [14], and [15], different document classification methods are proposed for a similar document classification problem as in this paper. Among them, Naive Bayes classifier [15] is shown to outperform other approaches, thus is adopted in our system to classify documents in the pruning process. We also improved their method by considering the concept hierarchy in ontology when computing the probabilities in Naive Bayes classifier.

The remainder of this paper is organized as the following. In Section 2, the terminology and metrics are formally defined. The methodology of our system is described in Section 3. Then the experiment results are presented in Section 4. Finally, we draw our conclusion in Section 5.

2   Preliminary

In this section, we formally defined some terminologies that will be used throughout the paper.

2.1   Document Similarity

Text documents are generally considered as bags of terms, and each document is represented in a multi-dimensional term vector. How to choose the set of feature terms to represent each document is discussed in the following section.

After obtaining the feature terms to represent documents, we construct a vector of real numbers for every document by assigning each feature term a numerical weight. The weight of a term is dependent on two factors: the importance of the term throughout all the documents and the strength of the term in a particular document. Therefore, the weight of term t consists of two parts: the global weight and the local weight. The global weight(gw) of a term t is defined as $gw(t) = \frac{|D|}{df(t)}$, where $|D|$ is the total number of documents in database, and $df(t)$ is the number of documents that contain term t. A definition of the local weight(lw) of a term t in a document d based on Poisson distribution [16] is given as below:

$$lw(t) = 1/(1 + exp(\alpha \times dlen) \times \gamma^{f(t,d)-1}) \tag{1}$$

where $\alpha = 0.0044$, $\gamma = 0.7$, $dlen$ is the length of document d, and $f(t,d)$ is the frequency of term t in document d.

For each feature term, the corresponding term weight in the vector is just the multiplication of the global weight and the local weight, i.e., $tw(t) = gw(t) \times lw(t)$.

After obtaining the term weight vector for each document, the similarity between two documents is defined as the cosine similarity of the two term weight vectors $\overrightarrow{tv1}$ and $\overrightarrow{tv2}$, given as below.

$$Sim(d1, d2) = Cosine(\overrightarrow{tv1}, \overrightarrow{tv2}) = \frac{\overrightarrow{tv1} \cdot \overrightarrow{tv2}}{\|\overrightarrow{tv1}\| \times \|\overrightarrow{tv2}\|} \tag{2}$$

2.2   Naive Bayes Classifier

The Naive Bayes classifier is commonly used for document classification. The basic idea of Naive Bayes classifier is to use the joint probabilities of words and classes to estimate the probabilities of classes given a document. In our system, a binary classification is conducted, i.e., documents are classified into two groups: either relevant or non-relevant. The binary Naive Bayes classifier is formally defined as the following: Given a set D of documents, the Naive Bayes classifier classifies a document d consisting of n feature words $(w_1, w_2, ..., w_n)$ as a member of the class

$$NB(d) = argmax(\hat{P}(c)\Pi_{i=1}^{n}\hat{P}(w_i|c)) \tag{3}$$

where $\hat{P}(c)$ is the estimated class probability, $\hat{P}(w_i|c)$ is the estimated joint probability of word t and class c, and $c \in \{0, 1\}$.

In the setting of this paper, a large number of positive example can be obtained, while only a very small number of negative examples exist. The reason is explained in a later section. In this case, the negative examples contribute little to the training set. The training set can be seen as consisting of only positive examples and some unlabeled examples. [15] proposed a Naive Bayes classifier on positive and unlabeled examples. The $\hat{P}(c)$ can be estimated on the training set by the following formulas:

$$\hat{P}(1) = \frac{|PD|}{|UD| + |PD|} \tag{4}$$

where $|PD|$ is the number of positive documents and $|UD|$ is the number of unlabeled documents. The negative class probability $\hat{P}(0)$ is estimated as $1 - \hat{P}(1)$.

The estimation of $\hat{P}(w_i|c)$ are given below:

$$\hat{P}(w_i|1) = \frac{N(w_i, PD)}{N(PD)} \tag{5}$$

$$\hat{P}(w_i|0) = \frac{Pr(w_i) - \hat{P}(w_i|1) \times \hat{P}(1)}{1 - \hat{P}(1)} \tag{6}$$

The word probability $Pr(w_i)$ is calculated by $\frac{N(w_i, UD)}{N(UD)}$, where $N(UD)$ and $N(PD)$ are the total number of words in unlabeled documents and positive documents. $N(w_i, UD)$ and $N(w_i, PD)$ are the occurrence of term w_i in unlabeled documents and positive documents, respectively. In order to make the probability estimation more robust with respect to infrequent words, smoothing methods can be used. Here, the classical Laplace smoothing [15] is conducted.

An example of estimating the conditional probabilities is as follows. Given 50 positive documents and 100 unlabeled documents, the positive class probability $P(1)$ is estimated as $50/150 = 1/3$, and the negative class probability $P(0)$ is $2/3$. For the feature term set containing six terms: $\{w_1, w_2, w_3, w_4, w_5, w_6\}$, their conditional probabilities in the positive class estimated from equation (5) are $\{0.1, 0.3, 0.15, 0.2, 0.1, 0.1\}$, and their conditional probabilities in the negative class estimated from equation (6) are $\{0.3, 0.7, 0.4, 0.1, 0.5, 0.6\}$. Then given an unlabeled document d_1 containing only feature terms w_2, w_3, w_5, w_6, the classifier given in equation (3) labels this document as negative because $P(1) \times 0.3 \times 0.15 \times 0.1 \times 0.1$ is less than $P(0) \times 0.7 \times 0.4 \times 0.5 \times 0.6$.

3   System and Methods

We have developed a prototype system to help users to retrieve useful biological literatures from a large amount of publications. The users will provide the keywords as the input and interact with the system during the retrieval process. In this prototype system, Gene Ontology is utilized as the background knowledge to organize documents, and the user feedbacks are used to refine the retrieved documents. Finally, the system returns a small set of documents that are considered as most relevant to users' preference. In this section, the methodology of our system is described and four main steps in the system are explained in details.

3.1   Pre-processing

During the pre-processing phase, Gene Ontology, which is originally described in a DAG (directed acyclic graph), is transformed to a tree hierarchy. If a term has multiple parents, it will have multiple instances in the transformed GO tree because it has different paths to the root term, which is important for the feature weighting discussed in a later section. For example, term "RNA transport" (GO:0050658) has two parent terms: "nucleic acid transport" (GO:0050657) and "establishment of RNA localization" (GO:0051236). Therefore, "RNA transport" has two instances in the transformed GO tree: one is at level 8 as a child of "nucleic acid transport", and the other one is also at level 6 as a child of "establishment of RNA localization". After the transformation of the Gene Ontology structure, the occurrences of GO terms are collected from the documents. All terms in three branches of Gene Ontology are searched for matching. When matching a subterm that is a substring of its parent GO term, both this term itself and its parent term are considered occurred. However, these duplicate countings are handled in the term weighting step as described in next subsection. The synonyms of GO terms defined in Gene Ontology are also considered equally as GO terms themselves. That is to say, if a synonym of a GO term appears in a document, the GO term is also considered occurred in the document. For instance, when searching for "peroxisome targeting sequence binding", "PTC binding" is also searched. By searching all documents, the number of occurrence of each GO term in each document is collected. Other statistical information is also collected at the same time, such as the length of every document, occurrence of every other word in each document. Non-informative words, such as "the", "we", are removed from the documents based on a stop-word list given by http://www.aifb.uni-karlsruhe.de/WBS/aho/clustering/stopwords.txt.

3.2   Feature Selection and Weighting

Traditionally documents are represented by a set of feature words. Feature selection is the process of selecting the set of words to represent documents. It benefits the clustering and classification by reducing the feature space and eliminating noisy features. In our system, the mutual information as defined in [13] is used as the criteria for feature selection. A thousand words of the most mutual information value throughout all the documents in each iteration are selected as the feature terms. Besides, a set of GO terms is also chosen as feature terms. A feature level is selected in the transformed GO tree, and all GO terms at this level serve as the feature terms.

The 1000 words of most mutual information and all the GO terms at the feature level in GO tree form the feature set. In our system, level 8 in Gene Ontology, which has the most number of GO terms in all levels, is selected as the feature level. Around 3000 GO terms at this level are chosen as feature terms to represent each document.

For those feature terms obtained by the most mutual information, a weight is assigned according to its global weight and local weight as defined in Section 2.1. A more complex weighting scheme is used for those feature terms from Gene Ontology. The original term weight computed from the above will be distributed and aggregated based on Gene Ontology structure. The weight of a term not at the feature level is distributed or aggregated to its ancestor or descendant terms at the feature level. If the term is at a lower level than the feature level, its weight is aggregated to all ancestors of this term in the feature level. If the term is in a higher level than the feature level, its weight is uniformly distributed to its children level by level until the feature level is reached.

Figure 1 illustrates an example of the distribution and aggregation process. A part of the Gene Ontology hierarchy is shown in Figure1. The two numbers beside each term at the feature level are the original weights computed for a document and the final weights after distribution and aggregation. If the second level in this figure is selected as the feature level, then only "Transport", "Secretion" and "Establishment of RNA localization" will serve as the feature terms when computing the document similarity. In this case, although the term "Establishment of RNA localization" never appears in the document (the original weight is 0), the weights of its children terms will be aggregated to the second

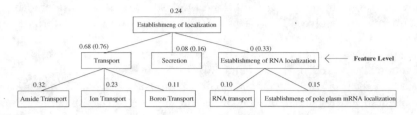

Fig. 1. Distribution and Aggregation of term weights

level. Therefore, term "Establishment of RNA localization" will gain weight of 0.25 from its children terms "RNA Transport" and "establishment of pole plasm mRNA localization". However, the weights of "Amide Transport", "Ion Transport" and "Boron Transport" are not aggregated to the second level, because their parent "Transport" is a substring of these chilren, and the occurrences of "Transport" has already been counted. Meanwhile, the weight of term "establishment of localization", which is located in the first level, is distributed uniformly to its children terms. Therefore, the final weight of feature terms "Transport", "Secretion" and "Establishment of RNA localization" in this document will be 0.76, 0.16 and 0.33 respectively.

3.3   Clustering and Representative Selection

Document clustering has been considered as an important tool for navigating large document collections. In our prototype system, after the user inputs the keywords to search, a set of documents is retrieved from the document corpora by exact keyword matching. In order to organize the documents in a meaningful way, in our system, the documents are clustered into groups according to their mutual similarities after each iteration. As described in the earlier section, rather than only considering the distribution of words in documents, our system compiles the background knowledge provided by biological lexicon into the similarity measurement.

Traditional document clustering methods only consider the distribution of words in documents, but ignore the fact that prior knowledge could be important in organizing the documents. Instead of measuring the document similarity directly by the distribution of words, our idea is to compile the background knowledge provided by biological lexicon into the similarity measurement, which is described in the earlier section.

In our system, Bi-Section-KMeans clustering method [11] is used, which has been shown to perform as good as other clustering algorithm, but much faster. If the number of documents is too large, a sampling technique is used to improve the clustering performance. A sample of all the documents is first clustered by Bi-Section-Kmeans algorithm, then the rest of the documents are assigned to their nearest clusters. After obtaining the document clusters, the centroid document of each cluster is chosen as the representative document. The user will only review the representative documents and rate each one as relevant or non-relevant.

In each iteration, documents are clustered and representatives are selected. The number of clusters is a parameter of the system and can be set by users. Users will read the representative documents and provide their evaluations. By looking at a small number of representatives in each iteration, users save significant amount of time from manually reading all search results.

3.4 Classification on Positive and Unlabeled Documents

In one iteration, the user will look at a small set of representative documents and label them. Based on their evaluations, the system will refine the document set and prune a set of documents that are considered as irrelevant to the user's interest. This pruning process is done by document classification. To be more specific, all the documents retrieved in this iteration are labeled as either "relevant" or "irrelevant" by a classifier.

The document cluster whose representative document is rated as "relevant" by users is called a positive cluster, and the document cluster whose representative document is rated as "irrelevant" by users is called a negative cluster. Since documents within a cluster are highly similar, documents in positive clusters can be seen as positive examples because their representative document is positively labeled by users. However, documents in the negative clusters can not be completely ruled out due to the coverage limitation of the representative document. According to this, documents in positive clusters are kept for the next iteration, while the documents in negative clusters are further classified. Thus, in the setting of the problem in this paper, the training set for the classifier contains only the positive examples (documents in positive clusters) and a set of unlabeled examples (documents in negative clusters), but no negative examples (The number of negatively rated representatives is small enough to be ignored).

In the Naive Bayes classifier defined in Section 2.2, word occurrence is an important factor affecting the class probabilities. In our system, the occurrences of GO terms are also distributed or aggregated in the same manner as the term weights in the clustering process. By doing this, even a GO term that appears in a document is not selected as the feature term, its probability of occurring in certain class is taken into account by its ancestor or descendant terms which serve as feature terms.

4 Experimental Results

A prototype search system is implemented in Perl based on the methodology proposed in this paper. One hundred thousand abstracts from PubMed, which are stored as plain text files in a 7200 rpm hardrive, are used to test our prototype system. These abstracts serve as the document universe in our experiments. In this section, experiment results are presented to demonstrate the effectiveness, efficiency and robustness of our proposed method. All the experiments below were run on a 2.4GHz P IV machine with 1GB RAM.

The following experimental method is conducted for evaluating the prototype system. First we search a set of key words, referred to as *reference keywords*, by exact keywords matching, then a set of documents are returned for this search query. This set of documents are considered as the truly relevant documents and serve as *the reference result set*. Then we introduce some noise in this set of keywords to generate the input keyword set for our prototype system. This can be done by removing some keywords from the reference keywords to generate a *reduced keyword set*. Naturally, the reduced keyword set will result in a larger result set than the reference keyword set by keyword matching. For example, we may choose "protein", "interaction network", "signaling", and "pathway" as the reference keywords. By searching these keywords via exact keyword matching, 1000 abstracts may be returned from all the 100,000 abstracts. After this, it is assumed that users only have partial knowledge on what they want to search. In this case, they may not know the complete set of keywords. For example, users may only provide three keywords: "protein", "interaction network", and "pathways". When using our prototype system, the users take these incomplete keywords as the input. The system first returns a larger set of documents that contain these reduced keywords, e.g., 2000 documents, then the system organizes these documents by clusters. In each iteration, users will look at the representative documents selected from these document clusters and evaluate them.

In our experiments, two criteria are used to evaluate the representative documents. One is asking biomedical researchers to provide their evaluations in each iteration, that is to say, biomedical researchers will use our prototype system to search for some literatures. The other criteria is to use a similarity threshold to automatically evaluate representatives. First the similarity between a representative and a set of documents is defined as the average similarity between this representative document and each document in the document set. Then, for each representative selected, the similarity between this representative and the documents returned by searching reference keywords is computed. A predefined threshold is used to distinguish relevant representatives and irrelevant representatives. automatically computes the similarities between all representative documents and the reference document set. If one representative's similarity to the correct documents is above the predefined threshold, this representative is considered as being rated "relevant". Otherwise, it is considered as an irrelevant representative. Thus, the system will be guided automatically without human intervention, then finishes the entire search process by only taking the keywords as input.

In our experiments, recall is used to evaluate the search performance of our prototype system. We denote the set of documents obtained by searching reference keywords as D_r and the set of documents our prototype system returns by taking the reduced keywords as input is denoted as D_o. The recall, which is defined as $\frac{|D_o \cap D_r|}{|D_r|}$, actually reveals how many truly relevant documents can be found be our iterative search approach.

A few parameters can be altered in our prototype system. The most critical one is the number of documents desired by users, that is, how many documents will be finally returned by the system. The larger this number is, the less number of iterations the system may run. However, the number of documents filtered out is essentially smaller. On the other hand, if this number is set too small, many truly relevant documents could be lost due to classification error. We will exploit how this parameter effects the response time and the recall of the system in this section.

The other parameter is the number of document clusters in each iteration. This parameter is actually the number of documents users need to examine in each iteration, because only one representative is selected for each cluster. Since we do not know the distribution of documents, this parameter has to be set based on experience. The presented experiment results show the robustness of our system against proper settings of cluster numbers.

The last parameter is the feature level selected in Gene Ontology. That is, terms in which level of Gene Ontology are used as feature terms. Intuitively, if the level is set too high or too low, the distinguishing power of terms in these levels is small because both high levels and low levels only contain a small number of terms. However, if the number of terms in the feature level is too large, the computational complexity of document similarity is essentially high, thus the system response time could be increased. We conducted an empirical study on how the selection of feature level effects the efficiency and performance of the system.

4.1   Effectiveness and Efficiency

The first reference keyword set we used to test is "metabolism", "expression", "regulation", "Phenotype", "protein", "mRNA" and "yeast", as shown in Table 1. By doing an exact keyword matching on this set of keywords, 310 documents are returned from our testing document universe. These 310 documents serve as the reference document set for performance evaluation.

In this experiment, the system was set to terminate when the number of remaining documents reaches half of the initial result document set. The number of document clusters was set to 10 in each iteration. Human users were asked to provide the evaluations of representatives. The time that users review the representative documents are not counted in the system response time. That is to say, the response time reflects the time for document clustering and classification.

Then we use the following three reduced keyword sets: "regulation, Phenotype and yeast", "metabolism, expression, regulation, Phenotype, protein and mRNA" and "regulation, mRNA and yeast" as the input keyword sets of our system. The combinations of these keywords are shown in Table 2.

Each of the three reduced keyword sets will result in thousands of documents by the exact keywords matching. Then our system began the iterative search process on these documents. Table 3 shows the number of iterations the system ran, the response time, and the recall on the three input sets. For example, when reduced keyword set 1 is used as the input, 3124 documents are returned by a keyword matching search. The system terminated after 4 iterations within 10 minutes, and the user reviewed 40 representatives, 18 of which were rated as

Table 1. Keywords

	Keyword Set 1	Keyword Set 2
keyword 1:	metabolism	protein
keyword 2:	expression	kinase
keyword 3:	regulation	enzyme
keyword 4:	Phenotype	synthetase
keyword 5:	protein	ligase
keyword 6:	mRNA	DNA
keyword 7:	yeast	

Table 2. Keyword set 1

	k1	k2	k3	k4	k5	k6	k7
Reference Set	X	X	X	X	X	X	X
Test Set 1			X	X			X
Test Set 2	X	X	X	X	X	X	
Test Set 3			X			X	X

relevant. About 1500 documents were filtered in this process, and 243 documents out of 310 documents of the reference result set are kept in the output, which indicates a recall of 78%.

Since the number of documents was reduced by half and a recall above 50% was achieved, the precision of returned results was also improved by our system. For instance, in this experiment, the precision of the results returned by searching the reduced keyword set 1 is $300/3000 = 10\%$. After 4 iterations, the system pruned 1500 documents, and kept 234 relevant documents, then the precision is $234/1500 = 16\%$. In order to test the relationship between recall and precision, we set the system to terminate after various number of iterations to generate result sets with different sizes, thus achieved different recall values. Figure 2 shows the precision values under different recall values. The results show that in order to include more relevant documents in the final results, larger number of documents should be kept as final.

A similar result is shown in Table 4 for the reference keywords set "protein", "kinase", "enzyme", "synthetase", "DNA" and "ligase". By searching for these reference keywords, 371 documents were obtained from the testing document universe. In this experiment, three different subsets of the reference keyword set. The results also show that our prototype system can identify over 70% of

Table 3. Performance on keyword set 1

	Iterations	Response Time	Recall
Test Set 1	4	600 s	78%
Test Set 2	5	645 s	75%
Test Set 3	4	570 s	72%

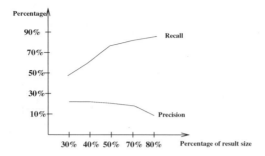

Fig. 2. Recall and Precision

the truly relevant documents while removing thousands of irrelevant documents in several iterations.

Table 4. Performance on keyword set 2

	Iterations	Response Time	Recall
Test Set 1	4	550 s	81%
Test Set 2	4	532 s	71%
Test Set 3	5	640 s	74%

Other keyword sets, such as "nucleotide binding, promoter, enzyme, expression and regulator", were also used as testing our system and generated similar results. The results demonstrate that by using our iterative approach combing with the background knowledge based document clustering and document classifications based on user feedbacks, the system can filter a large number of irrelevant documents in a tolerable time while still keep a large portion of useful documents.

4.2   Sensitivity Analysis

In this section, we analyze the sensitivity of our prototype system with respect to several parameters, i.e., the number of documents desired, the number clusters in each iteration, and the feature level selected in Gene Ontology.

We first test the system by varying the size of the initial document set and desired document set. The size of the initial document set is the number of documents obtained by searching the reduced keyword set. For testing the system against this parameter, we chose reduced keyword sets to make the number of returned documents range from 1000 to 10,000. Then the system worked on these documents and terminated after 50% of these documents were filtered. The size of the desired document set is the percentage of documents users want the system to return after interacting with the system. For this experiment, the percentage of output documents in the initial result document set was set to range from 10% to 90%. In this test, a complete reference keyword set was chosen to obtain 500 reference documents, while a reduced keyword set was chosen to obtain a fixed number of 5000 documents, which served as the initial document set. In addition, both objective human evaluation and the similarity threshold, which are explained at the beginning of this section, were used to evaluate the representatives. The results show the average performance of these two criteria.

Figure 3(a) shows the average response time of our system with increase of the size of initial document set. Since the sampling technique is used when the input document set is large, the major factor that increases the response time is the increase of iterations when the initial document set becomes large. In this sense, the overall system response time also partially depends on the user evaluations in each round, which affects the number of iterations. The effect of initial document set on the recall of the output is shown in Figure 3(b). The results demonstrate that the system performs robustly with the change of the size of initial document set.

In Figure 4(a) shows the system response time with the increase of size of desired document set. Similarly, the less percentage of documents users want, the more iterations the system runs, thus the larger the response time is.

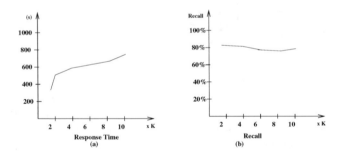

Fig. 3. Sensitivity to size of initial document set

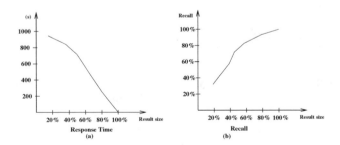

Fig. 4. Sensitivity to percentage of desired document set

Next, we tested the sensitivity of the prototype system with respect to the number of clusters in each iteration. In these experiments, a reference keyword set was chosen to result in 5,000 initial documents, and the system was set to terminate when half of these documents were filtered. Results show that the setting of cluster number, if not too low, actually has little power on affecting the efficiency and effectiveness of our system. However, this parameter can not be set too high, because this is actually the number of representative documents that users will review in each iteration. A tolerable setting of this parameter is from 4 to 10.

Finally, experiments were run for evaluating how the selection of feature level in Gene Ontology effects the system performance. The same system setting as in the previous experiment was used. The average response time and recall are shown in Figure 5.

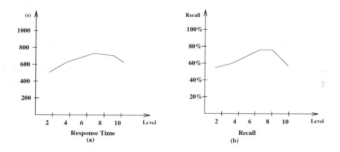

Fig. 5. Performance for different feature level selection

Note that level 8 of Gene Ontology has the most number of GO terms. The results confirm our analysis that the more feature terms are selected, the slower the system performs due to the increased complexity in computing document similarities. Also, the recall of system output is affected by the selection of feature level in Gene Ontology. The explanation of this is that too high or too low levels in Gene Ontology contain smaller number of terms, which limits the discriminating power of the feature space. Also, many branches of the Gene Ontology tree stop at a certain level, therefore, if a lower feature level is selected, the weights of terms in these branches can not be distributed to the feature level.

5   Conclusion

In this paper, a new iterative search paradigm is proposed. In our approach, document clustering is adopted to organize documents, and the user feedbacks are used to refine the retrieved documents. A new term weighting scheme is defined based on Gene Ontology structure, which benefits the document clustering by considering the hierarchy of biological concepts in the document similarity measurement. Naive Bayes classifier based on positive and unlabeled examples is used to classify documents via learning user feedbacks. By this approach, search results returned by search engines are organized in a meaningful way. Users review a much smaller number of representatives in each iteration. The system automatically filters a large number of irrelevant documents according to user feedback. A prototype biomedical literature search system has been built upon this iterative search paradigm. The experimental results show that the system can identify most of the truly relevant search results with the help of limited user interactions, while a large number of irrelevant results are pruned automatically during this process. Users can save a significant amount of time by only reviewing a small portion of search results. The performance of a search engine is improved by incorporating background knowledge and user feedbacks. Although a keyword-based search engine is used for testing purpose, our method can be easily adapted to other text search engines, e.g. eTBLAST [2].

References

1. PubMed (2005), available at http://www.ncbi.nlm.nih.gov/entrez/
2. eTBLAST, available at http://invention.swmed.edu/etblast/index.shtml
3. Brin S and Page L. (1998) The anatomy of a large-scale hypertextual web search engine, *WWW7 Conf., 1998*
4. Kleinberg J. (1998) Authoritative sources in a hyperlinked environment, *9th ACM-SIAM Symposium on Discrete Algorithms, 1998*
5. Suomela BP and Andrade MA (2005) Ranking the whole MEDLINE database according to a large training set using text indexing, *BMC Bioinformatics. Mar 2005*
6. Aphinyanaphongs Y., Tsamardinos I., Statnikov A., Hardin D., Aliferis CF (2005) Text categorization models for high quality article retrieval in internal medicine, *J Am Med Inform Assoc. 2005;12*
7. Vladimir D., David P., Mykola G., and Niall R. 2005 SOPHIA: An Interactive Cluster-Based Retrieval System for the OHSUMED Collection, *IEEE Transactions on Information Technology in Biomedicine, June 2005*
8. The Gene Ontology Consortium, available at http://www.geneontology.org.
9. Iliopoulos I., Enright AJ, Ouzounis CA (2001) Textquest: document clustering of Medline abstracts for concept discovery in molecular biology, *Pac Symp Biocomput, 2001*
10. Wilbur W. (2002) A thematic analysis of the AIDS literature. *PSB, 2002*
11. Andreas H., Steffen S., and Gerd S. (2003) Text Clustering Based on Background Knowledge, *Technical Report, 2003*
12. Yu H., Han J., and Chang K.C-C. (2002) PEBL: Positive Example-Based Learning for Web Page Classification Using SVM, *Proc. ACM SIGKDD, 2002*
13. Slonim N. and Tishby N. (2000) Document clustering using word clusters via the information bottleneck method, *ACM SIGIR 2000*
14. Liu B., Lee W.S., Yu P.S. and Li, X. (2002) Partially Supervised Classification of Text Documents, *Proc. 19th Intl. Conf. on Machine Learning, 2002*
15. Denis F., Gilleron R., and Tommasi M. (2002) Text classification from positive and unlabeled examples, *IPMU 2002*
16. Kim W., Aronson AR and Wilbur WJ. (2001) Automatic MeSH term assignment and quality assessment, *Proc AMIA Symp, 2001*

Automatic Annotation of Protein Functional Class from Sparse and Imbalanced Data Sets

Jaehee Jung[1] and Michael R. Thon[1,2]

[1] Department of Computer Science,
[2] Department of Plant Pathology & Microbiology,
Texas A&M University, College Station, TX, 77843 USA
{jaeheejung, mthon}@tamu.edu

Abstract. In recent years, high-throughput genome sequencing and sequence analysis technologies have created the need for automated annotation and analysis of large sets of genes. The Gene Ontology (GO) provides a common controlled vocabulary for describing gene function however the process for annotating proteins with GO terms is usually through a tedious manual curation process by trained professional annotators. With the wealth of genomic data that are now available, there is a need for accurate automated annotation methods. In this paper, we propose a method for automatically predicting GO terms for proteins by applying statistical pattern recognition techniques. We employ protein functional domains as features and learn independent Support Vector Machine classifiers for each GO term. This approach creates sparse data sets with highly imbalanced class distribution. We show that these problems can be overcome with standard feature and instance selection methods. We also present a meta-learning scheme that utilizes multiple SVMs trained for each GO term, resulting in improved overall performance than either SVM can achieve alone. The implementation of the tool is available at http://fcg.tamu.edu/AAPFC.

Keyword: Gene Annotation, Feature Selection, Gene Ontology, InterPro, Imbalanced Data.

1   Introduction

High-throughput genome sequencing and gene annotation methods have resulted in the availability or large sets of genes and predicted gene products (proteins) and to a large extent, the functions of many of these genes are still unknown, i.e. they are unannotated. Biologists deduce protein function through experimentation and as such, knowledge of gene function derived in this fashion is laborious and inexpensive. Given the wealth of genome data that are available now, one of the central problem facing researchers is the accurate prediction of protein function based on computationally obtained features of the proteins and the genes from which they are derived. Such computationally predicted functions are useful to guide laboratory experimentation and as an interim annotation, until protein function can be validated experimentally. Traditionally, protein function is expressed as free text descriptions but recently controlled vocabularies of various

M.M. Dalkilic, S. Kim, and J. Yang (Eds.): VDMB 2006, LNBI 4316, pp. 65–77, 2006.
© Springer-Verlag Berlin Heidelberg 2006

types have been employed. The Gene Ontology (GO) [22] provides a controlled vocabulary or terms for annotating proteins. In addition, the GO consortium describes the relationships among the terms with a directed acyclic graph (DAG), providing a rich framework for describing the function of proteins. GO terms are often assigned to proteins by teams of curators, who examine references in the scientific literature as well as features of the proteins. One of the central problems facing computational biologists is how to emulate this process.

As the need for GO annotation increases, various kinds of annotation systems are being developed for automated prediction of GO terms. Most methods rely on the identification of similar proteins in large databases of annotated proteins. GOtcha [12] utilizes properties of the protein sequence similarity search results (BLAST) such as the p-score, for predicting an association between the protein and a set of nodes in the GO graph. Several other recently described methods, including GoFigure [8], GOblet [5], and OntoBlast [19] depend on sequence similarity searches of large databases to obtain features that are used for predicting GO terms. These tools employ only BLAST results as attributes for prediction of GO terms, however, several systems utilize features besides BLAST search results. Vinayagam et el. [14,15] suggest a method to predict GO terms using SVM and feature sets including sequence similarity, frequency score the GO terms, GO term relationship between similar proteins. Al-shahib et el. [1] use amino acid composition, amino acid pair ratios, protein length, molecular weight, isoelectric point, hydropathy and aliphatic index as features for SVM classifiers to predict protein function. King et el. [9] employ not only sequence similarity, but also bio-chemical attributes such as molecular weight, and percentage amino acid content. Pavlidis et el. [13] predict gene function from heterogeneous data sets derived from DNA microarray hybridization experiments and phylogenetic profiles.

A number of different methods have been developed to identify and catalog protein families and functional domains which serve as useful resources for understanding protein function. The European Bioinformatics Institute (EBI) has created a federated database called InterPro (IPR) [23] which serves as a central reference for several protein family and functional domain databases, including Prosite, Prints, Pfam, Prodom, SMART, TIGRFams and PIR SuperFamily. InterPro families and functional domains are usually assigned to proteins using a variety of automated search tools. In addition, the InterPro consortium also maintains an InterPro to GO translation table that allows GO terms to be assigned to proteins automatically, on the basis of the protein domain content of the protein.

The availability of protein data sets annotated with GO terms and InterPro domains provides an opportunity to study the extent to which InterPro can be used to predict GO terms. The InterPro database contains over 12,000 entries and the GO contains over 19,000 but proteins are usually annotated with a few terms from each database, resulting in a sparse data set. In addition, a large set of proteins will contain only a few positive examples of each GO term, leading to extremely biased class distribution in which less than 1% of the training instances represent positive examples of a GO term.

Many studies have shown that standard classification algorithms perform poorly with imbalanced class distribution [7,10,16]. The most common method to overcome this problem is through re-sampling of the data to form a balanced data set. Re-sampling methods may under-sample the majority class, over-sample the minority class, or use a combination of both approaches. A potential drawback of under-sampling is that effective instances can be ignored. Over-sampling, however, is not without its problems. The most common approach is to duplicate instances from the minority class but Ling et el. [11] show that often times this approach does not offer significant improvements in performance of the classifier, as compared to the imbalanced data set. The other approach is the Synthetic Minority Over-sampling Technique (SMOTE) [2], which is an over-sampling technique with replacement in which new synthetic instances are created, rather than simply duplicating existing instances. Under-sampling can potentially be used to avoid the problems of over-sampling [10,20]. Under-sampling removes instances from the majority class to create a smaller, balanced data set. While other approaches such as feature weighting can be employed, under-sampling has the added benefit of reducing the number of training instances that are required for training, thus reducing the difficulties of training pattern recognition algorithms on very large data sets.

In this paper we consider the application of statistical pattern recognition techniques to classify proteins with GO terms, using InterPro terms as the feature set. We show that many of the problems associated with sparse and imbalance data sets can be overcome with standard feature and instance selection methods. Feature selection in an extremely sparse feature space can produce instances that lack any positive features, leading to a subset of identical instances in the majority class. By selectively removing these duplicated instances, or keeping them, we trained two SVMs that have different performance characteristics. We describe a meta-learning scheme that combines both models, resulting in improved performance than can be obtained by using either SVM alone.

2   Methods

2.1   Dataset

The data set used for this study was comprised of 4590 annotated proteins from the *Saccharomyces cerevisiae* (Yeast) genome obtained from the UniProt database [26]. This protein contains manually curated GO annotations as well as InterPro terms automatically assigned with InterProScan.

The data set contains 2602 InterPro terms and 2714 GO terms with an average of 2.06 InterPro terms and 3.99 GO terms assigned to each protein. Table 1 illustrates the imbalanced nature of the data set. In this study, each GO term was considered as an independent binary classification problem and therefore, all proteins annotated with a GO term are treated as positive instances (GO+)

and the remaining proteins treated as negative instances (GO-), resulting in highly biased class and feature distribution. For the purpose of this study, we only considered data sets that contained at least 10 GO+ proteins.Therefore, proteins annotated only with GO terms that did not meet this criterion were removed from the data set, resulting in a reduction of the size of the data set to 4347 proteins.

Table 1. Examples of randomly selected classes (GO terms) and features (InterPro terms) illustrating the imbalanced and sparse nature of the data set

GO term	Number of Positive Examples	Number of Negative Examples	InterPro term	Number of Positive Examples	Number of Negative Examples
GO:0000001	22	4568	IPR000002	5	4585
GO:0000022	15	4575	IPR000009	2	4588
GO:0000776	12	4578	IPR000073	13	4577
GO:0005635	35	4555	IPR000120	2	4588

2.2 Under-Sampling

Several methods are available for creating balanced data sets. If the features are continuous, we can perform over-sampling using methods such as SMOTE [2] which created new interpolated value for each new instance. In our case, however, the data set is binary format so this method cannot be used. In most cases under-sampling is considered to be better than over-sampling in terms of changing in misclassification costs and class distribution [2]. Another issue about the under-sampling is how ratio positive verse negative to make balanced set is optimized for training. In the point of the dealing with the imbalanced data problem, Al-shahib et el. [1] applied various under-sampling rates from 0% to 100% and conclude that the fully balanced set which have same number of positives and negatives, give the best performance. In light of this prior work, we performed under-sampling to create fully balances data sets for each GO term.

For each data set, we performed under-sampling of the majority class (GO-negative proteins) to create a balanced data set for SVM induction. We compared the performance of four under-sampling methods: Farthest, Nearest, Cluster and Random . In the first two cases, we used Euclidean distance, computed on the basis of the InterPro terms content of each protein as a measure of distance. The Farthest and Nearest methods select proteins from the negative class that have the greatest and least distance from the positive class respectively. The Cluster method first performs hierarchical clustering of the negative class where the number of clusters formed equals the number of instances in the positive class. A single protein from each cluster is selected randomly. The Random method randomly selects proteins from the negative class.

Let D_{All} be the set of all of IPR and GO data. We define the example of data set as $D_{All} = \{(X_i, Y_j) \mid i, j=1, \cdots, k \}$, where k is the number of proteins, and $X=(x_1, x_2, \cdots, x_l) \in \text{IPR}\{0,1\}$ are feature vectors and l is the number of InterPro (IPR) features in the data set. $Y=(y_1, y_2, \cdots, y_m) \in \text{GO}=\{0,1\}$ is the class designation (GO terms) and m is the number of GO terms in the data set.

2.3 Feature Selection

We employed four different objective functions for feature selection: chi-squared ($\chi 2$), information gain, symmetrical uncertainty, and correlation coefficient.

Classical linear correlation [18] measures the degree of correlation between features and classes, and ranges from -1 to 1. If the features and the class are totally independent, then the correlation coefficient is 0. The traditional linear correlation method is very simple to calculate, but it assumes that there is a linear relationship between the class and the feature, which is not always true [18]. To overcome this shortcoming, the other correlation measures based on the theoretical concept of entropy were also assessed for feature selection. Information gain is a measurement based on entropy, and measures the number of bits of information obtained for class prediction [17]. However, information gain have non-normalized value and it is biased toward of feature with more value. To compensate for this disadvantage, symmetrical uncertainty value is normalized from 0 to 1 and un-biased in terms of feature content. The idea of symmetrical uncertainty is based on the information gain, but applied value is normalized and un-biased toward feature with more value [18]. When calculating the contingency between features and a class of interest, the $\chi 2$ statistic measures the lack of independence. As the $\chi 2$ statistic values increases, the dependency between features and classes also increases [17,18].

The features were ranked using each of the objective values and a sequential forward selection search algorithm was used for feature selection. Forward selection was used since it is considered to be computationally more efficient than backward elimination [4]. The feature inclusion threshold for 12 randomly selected data sets was determined by computing the error rate during each stepwise feature addition and finding the minimal error rate. The average threshold value for the 12 data sets was used for the remaining data sets.

2.4 Implementation

Feature selection experiments were performed with WEKA [27]. Under-sampling and SVM induction were performed with MATLAB [24] using the pattern recognition toolbox [25].

3 Experiments

Individual data sets are constructed for each GO term which are then subjected to feature selection and instance selection prior to model induction. We performed preliminary experiments using a variety of machine learning algorithms to determine which one produced the best performance with our data set. For example, preliminary experiments showed that among data sets with more than 100 positive GO terms, the average sensitivity obtained from SVM and decision tree models was almost identical, but the specificity and accuracy of SVM models was higher (0.82 and 0.93 respectively) than values obtained from decision tree models (0.70 and 0.5 respectively). When the number of positive GO terms were small, only slight differences in sensitivity and specificity between SVM and decision tree were obtained, but the accuracy of SVM was consistently better than decision tree. Therefore, SVM was chosen for further experiments.

Because of the extremely sparse nature of the data set, the feature selection step can remove all InterPro terms from some proteins, resulting in proteins that completely lack features. In most cases, feature selection resulted in a large number of GO- proteins in each data set. We theorized that such a large number of redundant proteins in the data sets could lead to skewed performance of the SVM so for each GO term, we constructed two data sets. *Model 1* refers to the SVM learned from the data set containing the redundant GO- proteins and *Model 2* refers to a smaller set in which redundant proteins were removed prior to model induction (Fig. 1). We expected that *Model 2* would result in SVM with higher accuracy than only *Model 1*.

3.1   Feature and Instance Selection

We randomly selected 50 *Model 1* data sets and compared the performance of the feature selection and instance selection methods. The relative performance of the various methods were compared using error rate and AUC by 10-fold cross validation. The chi-squared method outperformed the other feature selection methods (Table 2) and was used to prepare data sets for instance selection. The Farthest method provided the best instance selection performance (Table 3) and was selected to create balanced data sets for SVM induction.

Table 2. Performance comparison of 4 different feature selection methods. Features were ranked using one of four objective functions (SU: symmetrical uncertainty, INFO: information gain, CHI: chi-squared, ABS: absolute correlation coefficient) and sequential forward selection was performed to optimize. Values represent average over 50 data sets.

Method	Sensitivity	Specificity	AUC	Error Rate
SU	0.98	0.83	0.73	0.01
CHI	0.99	0.93	0.87	0.01
INFO	0.78	0.98	0.85	0.12
ABS	0.77	0.25	0.79	0.43

We used 10-fold cross validation to compare the performance of SVMs trained using *Model 1* and *Model 2*. In general, *Model 1* SVMs had very low false

Table 3. Performance comparison of 4 different under-sampling methods. Nearest is the result of applying for choosing negative instances as the nearest method, and Farthest is the farthest negative instances. Cluster is clustering and choosing randomly negatives. Random is randomly selected.

Method	Sensitivity	Specificity	AUC	Error Rate
Farthest	0.94	0.94	0.78	0.03
Nearest	0.73	0.79	0.74	0.52
Cluster	0.90	0.90	0.77	0.09
Random	0.87	0.93	0.77	0.08

negative rates but had high false positive rates whereas **Model 2** SVMs tended to have lower false positive rates (Fig. 2). On average, **Model 1** has 0.32 false negative instances per SVM but 297.54 false positive instances and 4024 true negative instances per SVM among 4347 proteins. Of the 374 SVMs trained, 84% have less than 1 false negative instance using **Model 1** (Fig. 2(a)). Therefore, we conclude that this model is effective at classifying positive instances, although it should be noted that **Model 1** trained SVMs have high false positive rates. Since properties of both models were desirable for our classifier, we developed a meta-learning scheme that incorporated both models and includes a final filtering step, in order to reduce the false positive rate.

3.2   Test Procedure

Data flow for the prediction step is shown Fig. 3. We focus on keeping the true positive rate as high as possible so **Model 1** is utilized as first step. The **Model 1** classifier plays a role of excluding the most negative instances, but has the risk of making false positive classifications. Proteins classified as positive by **Model 1** are classified again using **Model 2**, thereby reducing the number false positive proteins from 297.54 to 110.63 on average.

The third step is comprised of a decision rule that we devised based on observations we made of the dataset. Under the assumption that a positive relationship exists between GO terms and InterPro terms, we define the following decision rule: For each GO term assigned to a protein, we identify whether a training proteins exists with that GO term and an InterPro term assigned to the predicted protein. If at least one association exists, the the predicted GO term is retained, otherwise it is removed from the set of predicted GO terms.

We compared the precision of the suggested classification procedure (Fig. 3 Process B) with the precision of **Model 1** alone (Process A), where precision is measured as the number of true positive GO terms divided by the number of predicted GO terms. We randomly selected and held out 435 proteins from the training data set to use for comparative analysis of the two classification procedures. The average precision of Process A, which applies only **Model 1**, is 0.3411, while Process B which applies both SVM models is 0.3523.

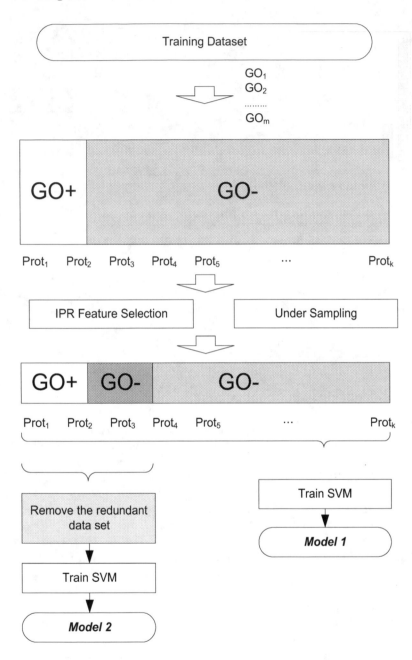

Fig. 1. Flow chart for the training process. For each GO term, the data set is reduced by feature selection and under sampling, which are performed independently using the full data set. SVM *Model 1* is trained on the reduced data set. To train *Model 2* the data set is further reduced, by removing redundant negative instances (see text for details).

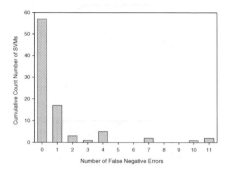 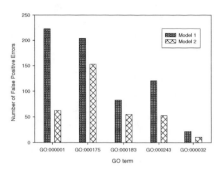

Fig. 2. Performance of SVMs estimated by 10-fold cross validation. (a) Cumulative count of number of false negative errors produced by 374 independent SVMs (b) False positive errors produced by *Model 1* and *Model 2* for five randomly selected GO terms .

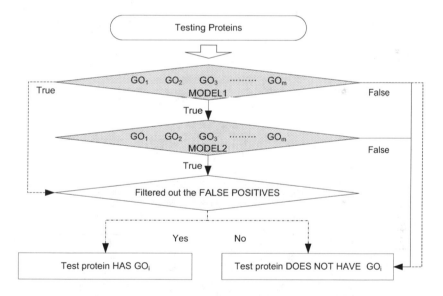

Fig. 3. Flow chart for the testing process. The dotted line represents use of *Model 1* only (Process A). The solid line represents use of both *Model 1* and *Model 2* (Process B). The filtering step is used in both cases.

3.3   Comparison to Other Methods

Using the training set, we prepared SVMs for each GO term. The precision is employed again as a metric to compare the performance of our method to that of other described methods. Most automated GO annotation methods are only available as web-based forms designed to process one protein at a time. Therefore

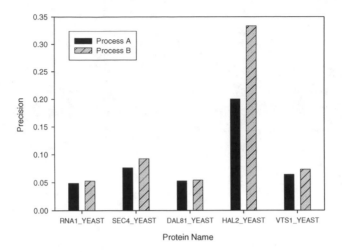

Fig. 4. The precision of two process: Process A (apply *Model 1* and filtered out), Process B (apply *Model 2* and filtered out)

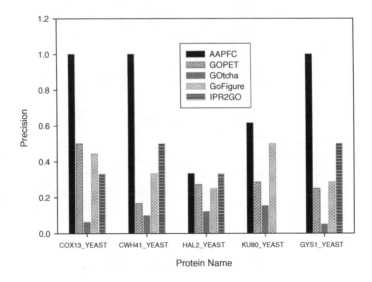

Fig. 5. Comparison of the precision proposed classification method (AAPFC) to four other methods by randomly selected 5 proteins, which does not include in the known training set in the *S. cerevisiae*.

we randomly selected nine proteins from the hold out set to use for comparison to other published annotation methods. One exception to this is IPR2GO, which is a manually curated mapping of InterPro terms to GO terms that is maintained by the InterPro consortium. We implemented a classifier using the IPR2GO rules and estimated performance using 10-fold cross validation. Our method, which we term Automatic Annotation of Protein Functional Class (AAPFC) includes trained SVMs for every GO term in which ten or more positive protein instances could be found in the training data set. In this comparison, AAPFC had a precision of 0.3241 while that of IPR2GO was 0.1777. Additionally, the precision of three other GO annotation tools was, in most cases considerable lower than AAPFC (Fig. 5). We used the author recommended confidence thresholds of 20% and 50% for the GOtcha and GOPET methods, respectively, and employed an e-value cutoff of 1e-5 for GoFigure. On average, precision is 0.53 for AAPFC, 0.17 for GOPET, 0.05 for GOtcha, 0.29 in GoFigure, and 0.20 in IPR2GO. Surprisingly, AAPFC outperformed IPR2GO, suggesting that there are many protein functions that can be predicted from InterPro terms, that cannot be described an a simple one-to-one translation table.

4   Conclusions

In this paper, we propose a method for assigning GO terms to proteins using InterPro terms as features and learning independent SVMs for each GO term. By creating two data sets, each having different properties, and learning two SVMs for each GO term, we developed a meta-learning scheme that benefits from the strengths of each model. Our long-term plans are to develop a system for assigning GO terms to proteins using multiple feature types, including biochemical properties (amino acid content, etc.), phylogenetic profile, sequence similarity, and others. Our current strategy treats each GO term as an independent learning problem. This has some practical benefits in that individual classifiers or sets of classifiers could be learned or re-learned over time without the need to re-learn the whole system. On the other hand, this approach assumes that all GO terms are independent. During our initial steps of data exploration and data cleaning, we observed a high correlation coefficient among some pairwise GO terms, indicating that there is dependency among some GO terms. Therefore, future work we propose to utilize stacked generalization as an approach to capture dependence among GO terms into the learning method. The outputs of the classifiers described here can be used as inputs to another classifier, thus enabling the dependence among GO terms to be utilized for classification.

References

1. Al-shahib, A., Breitling, R. and Gilbert, D. : Feature Selection and the Class Imbalance Problem in Predict Protein Function form sequence. Applied Bioinformatics Vol.4 (2005) 195-203
2. Chawla, N.V., Bowyer, K. and Hall, L.O. : Kegelmeyer,W.P. SMOTE: Synthetic minority over sampling technique. Journal of artificial Intelligence Research , Vol.16 (2002) 321-357

3. Drummond, C. and Holte, R.C. : C4.5,Class Imbalance, and Cost sensitivity: Why Under-sampling beats Oversampling. ICML'2003 Workshop on Learning from Imbalanced Datasets II. (2003)
4. Guyon, I. and Elisseeff, A. : An introduction to variable and feature selection. Journal of Machine Learning Research Vol.3 (2003) 1157-1182
5. Hennig, S., Groth, D. and Lehrach, H. : Automated Gene Ontology annotation for anonymous sequence data. Nucleic acids Research (2003) 3712-3715
6. Huang, J., Lu, J. and Ling, C.X. : Comparing Naive Bayes ,Decision Trees, and SVM using Accuracy and AUC. Proc. of The Third IEEE Inter. Conf. on Data Mining (ICDM) (2003) 553-556
7. Japkowics, N. and Stepen, S. : The class imbalanced problem : A systematic study. Intelligent Data Analysis Vol.6 (2002)
8. Khan, S., Situ, G., Decker,K. and Schmidt,C.J. : GoFigure:Automated Gene Ontology annotation. Bioinformatics Vol.19 (2003)
9. King, R.D., Karwath, A.,Clare,A. and Dephaspe,L. : Genome scale prediction of protein functional class from sequence using data mining. Proc. of the sixth ACM SIGKDD Inter. Conf. on Knowledge discovery and data mining (2003)
10. Kubat, M. and Matwin, S. : Addressing the curse of Imbalanced Training sets : One-sided Selection. Proc. of the Fourteenth Inter. Conf. on Machine Learning Proc. (ICML) (1997) 179-186
11. Ling, C. and Li, C. : Data mining for direct marketing :problem and solution. Proc. of the Fourth Inter. Conf. on Knowledges Discovery and Data Mining(KDD) (1998) 73-79
12. Martin, D.M., Berriman, M. and Barton, G.J. : GOtcha : A new method for prediction of protein function assessed by the annotation of sever genomes. BMC bioinformatics Vol.5 (2004)
13. Pavalidis, P., Weston, J., Cai, J. and Grundy, W.B. : Gene Functional Classification From Heterogeneous Data. Proc. of the Fifth Inter. Conf. on Research in Computational Molecular Biology (RECOMB) (2001) 249-255
14. Vinayagam, A., Konig, R., Moormann, J., Schubert, F., Elis, R. , Glatting, K.H. and Suhai, S. : Applying support vector machine for gene ontology based gene function prediction. BMC Bioinformatics Vol.19 (2003)
15. Vinayagam, A., Val, C.D, Schubert, F., Elis, R., Glatting, K.H., Suhai, S. and Konig, R. : GOPET : A tool for automated predictions of Gene Ontology terms. BMC Bioinformatics Vol.7 (2006)
16. Weiss,G.M. : Mining with rarity : A unifying framework. ACM SIGKDD Explorations Newsletter Vol.6 (2004) 7-19
17. Yang, Y. and Pedersen, J.O. : A comparative study on feature selection in text categorization. Proc. of the Fourteenth Inter. Conf. on Machine Learning (ICML) (1997) 412-420
18. Yu, L. and Liu, H. : Feature Selection for high-Dimensional Data: A Fast Correlation-based filter solution. Proc. of the Twentieth Inter. Conf. on Machine Learning (ICML) (2003)
19. Zehetner, G. : OntoBlast function: from sequence similarities directly to potential functional annotations by ontology terms. Nucleic acids Research (2003) 3799-3803
20. Zhang, J. and Mani, I. : kNN Approach to Unbalanced Data Distributions: A case study involving Information Extraction. ICML'2003 Workshop on learning from imbalanced datasets II (2003)
21. Zheng, Z., Wu, X. and Shrihari, R. : Feature selection for text categorization on imbalanced data. ACM SIGKDD Exploration Newsletter Vol.6 (2004) 80-89

22. Gene Ontology(GO) Consortium, http://www.geneontology.org/
23. InterPro, http://www.ebi.ac.uk/interpro/
24. MATLAB, http://www.mathworks.com/
25. Pattern Recognition Toolbox for MATLAB,
    http://cmp.felk.cvut.cz/~xfrancv/stprtool/
26. UniProt, www.uniprot.org/
27. WEKA, http://www.cs.waikato.ac.nz/~ml/

Bioinformatics Data Source Integration Based on Semantic Relationships Across Species

Badr Al-Daihani[1], Alex Gray[1], and Peter Kille[2]

[1] School of Computer Sciences, Cardiff University
5 The Parade
Cardiff, CF24 3AA
UK
{badr,a.w.gray}@cs.cf.ac.uk
[2] School of Biosciences, Cardiff University, Museum Avenue,
Cardiff, CF10 3US, UK
kille@cardiff.ac.uk

Abstract. Bioinformatics databases are heterogeneous, differ in their representation as well as in their query capabilities across diverse information held in distributed autonomous resources. Current approaches to integrating heterogeneous bioinformatics data sources are based on one of a: common field, ontology or cross-reference. In this paper we investigate the use of semantic relationships across species to link, integrate and annotate genes from publicly available data sources and a novel Soft Link approach is introduced, to link information across species held in biological databases, through providing a flexible method of joining related information from different databases, including non-bioinformatics databases. A measure of relationship closeness will afford a biologist a new tool in their repertoire for analysis. Soft Links are identified as interrelated concepts and can be used to create a rich set of possible relation types supporting the investigation of alternative hypothesis.

1 Introduction

One of the largest obstacles to bioinformatics data source integration is that many communities and laboratories use different, proprietary technologies that inhibit interoperability. Available bioinformatics data integration systems are concerned with unifying data that are related semantically, across heterogeneous data sources. Most data sources are centred on one primary class of objects, such as gene, protein or DNA sequences. This means that each database system has different pieces of biological information and knowledge, and can answer queries addressed to its individual domain, but cannot help with queries which cross domain boundaries. Many bio-informatics databases do not provide explicit links to data held in other databases, such as, orthology and other types of relationships. Relationships between data held in such resources are usually numerous, and only partially explicit. When these relationships are expressed as links, or predefined cross-references, they suffer numerous problems – such as vulnerability to naming and value conflicts. These

M.M. Dalkilic, S. Kim, and J. Yang (Eds.): VDMB 2006, LNBI 4316, pp. 78–93, 2006.

cross-references are created by analysis. Typically such cross-references are stored as the accession number of the entries in the linked databases. Links are added to data en-tries for many reasons: e.g. biologists insert them when they discover a confident relationship between items, or data curators insert them as a sign of structural relationships between two databases. They are usually stored as a pair of values, for example, target-database/accession number, represented as a hyperlink on a webpage (Bleiholder, Lacroix et al. 2004; Leser and Naumann 2005). In existing integration systems, joining information held in different data sources is based on the uniqueness of common fields in the sources. This paper presents a novel approach for providing a flexible linkage between sources, using a soft link model. A Soft Link Model (SLM) can also link suitable data sources of each species that have an orthology, common function, sequence similarity, GO term, or any other relationships. These links are generated using the similarity between values of text fields, or based on sequence homology, or orthology between sequence fields. It also provides a means for integrating data from other disciplines such as biology, chemistry, pharmaceutical research and drugs. This allows integration of genome data with other related data, such as metabolic information and structural biological data (Robbins 1995). This assists biologists to investigate different DNA structures, to foresee sequence functions and to identify the position of a gene of interest in for instance mouse, fly or Wormbase genomes, depending on the specifics of the research questions. This paper consists of: related work in Section 2, Section 3 the problem, Section 4 the novel approach to linkage, Section 5 demonstrates how it can work, Section 6 Section, Section 7 discussion and Section 8 conclusions.

2 Related Work

Data integration by linkage of bioinformatics data sources has been attracting re-search attentions for several years. Many different approaches have been proposed. Existing systems use a number of different approaches. Currently, there are four basic models of data integration:

A) Mediation offers an integrated view of data sources through wrappers. A mediator provides a virtual view of the integrated sources. It interacts with data sources via wrappers, and handles a user query by splitting it into sub-queries, sending the sub-queries to appropriate wrappers and integrating the results locally to provide answers to queries. Examples of mediation systems are covered in (Letovsky 1999; Gupta, Ludäscher et al. 2000; Freier, Hofestadt et al. 2002; Lacroix and Critchlow 2003)[10-25]

B) Federation In a federation the database systems are independent. Data continues to be stored in its original location and is retrieved via a middleware component, which uses a common data model and mapping scheme to map heterogeneous data source schemas into the target schema. Maintenance of a common schema can be costly due to frequent changes in data source schemas. K2/Biokleisli (Lacroix and Critchlow 2003) and DiscoveryLink (Lacroix and Critchlow 2003) are examples of this ap-proach.

C) Data Warehouses bring data from different data sources into a centralized local system using a global schema. Data warehouses often use wrappers to import data from remote sources. These data are materialized locally through a global schema used to process queries. While this simplifies the access and analysis of data stored in heterogeneous data repositories, the challenge is to keep the data in the warehouse current when changes are made to the sources, this is especially problematical when the warehouse is large. It requires substantial maintenance effort, lacks scalability and requires a deep understanding of data schemas. Thus the cost of maintenance, storage and updating data is high. Data can however be readily accessed, without delay or bandwidth limitation, and duplication, errors and semantic inconsistencies are removed through the warehousing of data. GUS (2004), AllGenes(Lacroix and Critchlow 2003) are examples of this approach

D) Link-based Integration. Many databases provide hypertext links to other data sources. Usually accession numbers (AC) or global identifiers are used for interlinking. Some databases use other attributes for interlinking, such as GO terms (Ashburner, Ball et al. 2000), EC numbers and CAS registry numbers (Buntrock 2001). However, as different databases use different identifiers for terms with equivalent entries, this approach is labour intensive. For this reason most databases only provide links to the most relevant databases via accession numbers (Kohler 2003; Kohler 2004). Examples include: the Sequential Retrieval System (SRS)(Etzold, Ulyanov et al. 1996) (Lacroix and Critchlow 2003), BioNavigator and Entrez (Maglott, Ostell et al. 2005). SRS is based on indexing flat files and relies on explicit links between sources. These models suffer from the following: i) the type of linkages are fixed and difficult to change as they are determined by wrappers in A, the middleware component in B, the code that executes the warehousing in C. ii) breaking of links. This occurs, when a URL changes. It primarily affects D, but can occur in A and B. iii) changes in the Databases (DB) are delayed, thus the data being analyzed is not current. This only affects data warehousing. iv) Difficulties in linking to non-bioinformatics DB. D can handle this in a limited manner. v) Inability to sup-port multiple types of relationship; all four are subject to this limitation. In the SLM we address these issues so that researcher requiring a more flexible approach which operates on the most recent version of data will be supported. The main contributions of this paper are: (a) a framework for defining, describing and supporting the integration of local experimental data with public data sources; (b) packaged tools and soft-ware to create solutions for integration.

3   Problem Definition

Many strategies are available to join data across heterogeneous data sources. These strategies suffer from several problems.

3.1   Join Based on Match Fields Values (Keyword-Based)

The most familiar approach for integrating data is to match fields between data sources. Two entries from diverse sources can be integrated using the identity of an accession number. However, different sources may use different terminology. For example, one source may use scientific name for a species (Mus musculus or Es-cherichia coli) while others use the common name (mouse or Bacterium coli). Since there are often no common keys to join tables and databases keyword-based does not always work.

3.2 Usage of Ontology (Concept-Based)

An Ontology is the formal specification of the vocabularies of concepts and the relationships among them for a discipline. Ontologies have played a key role in describing semantics of data in data sources. Using an ontology in data source integration has been studied by for example (Collet, Huhns et al. 1991; Decker, Erdmann et al. 1999). This often involves mapping the data semantically to a proprietary ontology. This supports the integration of data from different external data sources in a trans-parent way, capturing the exact proposed semantics of the data source terms, and removing mistaken synonyms. Systems like TAMBIS and OBSERVER allow users to create their query over an ontology without the need to be aware of the data sources' representation or to have direct access to these sources. An ontology can help in solving the interoperability problems among heterogeneous databases, since it establishes a joint terminology between different research communities. A variety of ontologies and approaches have been used in the bioinformatics domain, which mean sources have a different view, for example different levels of granularity and finding the minimal ontology commitment(Gruber 1995) becomes a hard task. However, onotlogies can resolve semantic heterogeneity problems, broaden the scope of searches on the integrated data sources; and enhance the quality and integrity of the data to be integrated from the heterogeneous sources. There are some limitations to their use. First, ontologies are incomplete in their representation of a domain, since usually data about bioinformatics sources is either non-existent or too specific. Secondly, the computational tools which compute mapping between data in sources and Ontology concepts are still immature and might be difficult to apply effectively.(Carel 2003). Moreover, the lack of a common vocabulary makes it hard to compare different source ontologies.(Wache, ogele et al. 2001). Also, since the understanding of biological systems keeps changing, and the technical domains crossed by genomics and bioinformatics are disparate, there are difficulties in capturing all the information from biological systems.(Venkatesh and Harlow 2002). Thus there will be a tendency for the most specific ontologies to differ and become less interoperable.

3.3 Cross-Referencing (Hard Links)

Another integrating strategy is the use of hard links. In this approach the user begins his/her query with one data source and then follows hypertext links to related information in other data sources (Baxevanis and Ouellette 2001). These relationships are expressed as links, or predefined cross-references. Typically, such cross-references are stored in tables as the accession numbers of entries in the databases. These cross-reference tables are then used for inter-operability. Links or cross-references are determined in several ways: computation of similarity between sequences using alignment tools, for instance BLAST, or mining literature(Leser and Naumann 2005). (Bleiholder, Lacroix et al. 2004) define how links are added to data entries in bioinformatics data sources:

- Researchers add them when discovering a confident relationship between items.
- Data curators add them as a sign of structural relationships between two data sources.
- Computational tools, for instance BLAST, add them when discovering similarity-ties between two data entries.

However, cross-references, or hard links, have several drawbacks: they are subject to naming and value conflicts. For example, if a curator changes or deletes an entry which is related to an entry in another data source, the link fails(Bleiholder, Lacroix et al. 2004; Leser and Naumann 2005). Moreover these links are syntactically poor because they are present only at a high granular level: the data entry. In addition, they are semantically poor because they do not provide any explicit meaning other than the data entries are in some way related (Bleiholder, Lacroix et al. 2004)

In summary, the integration challenge is that source data cannot be joined using simple term-matching or comparison operators. Even more sophisticated approaches which use ontologies to enumerate joinable terms (Kashyap and Sheth 1996) are often not sufficient. (Gupta, Ludäscher et al. 2000). Instead a join should be performed based on whether there is a semantic relationship between objects.

3.4 Join Based on Relationships (Semantic Relationship)

We suggest semantic relationships between concepts' properties may solve these problems. Thus a join is made where there is a semantic relationship between objects. We describe this approach in (Al-Daihani, Gray et al. 2005). Here we describe it in some detail after defining the terms used. Due to the lack of a systematic approach to linking databases, and because the number of existing biological databases is too high to survey, there is a poor or low degree of interlinking across such data-bases(Williams 1997). Biologists who are interested in a gene or protein of one species may find it extremely difficult to find related, relevant data on the same gene in databases for another species. This barrier exponentially increases if the biologists require relevant cross species information about 10 or more of genes identified by genomic experimentation. The relationships that exist between biological objects are important factors in determining which bioinformatics data sources should be linked. A relationship which provides flexible and loosely coupled linkages will allow biologists to discover related data across the biological universe and is the aim of our soft linkage. Soft Links are modelled via concepts whose interrelationship is detected using a rich set of possible relation types.

4 Soft Link Model

4.1 Definitions

Definition 1: Relationship Closeness (RC) is a measure of the closeness of concepts, where 'closeness' is defined in terms of different dimensions. This measure indicates the degree of closeness, i.e, it is a measure of how close one concept is to another. It is expressed as a percentage, with 100% meaning C1 is the same as C2. A high value of RC indicates there is a significant link between the concepts, and a low value of RC indicates there is no link or no significant link between the concepts.

Definition 2: C is a set of concepts. $C = \{c_1, c_2, \ldots \ldots c_n\}$, where the concept, c_i , is an entry in a database that represents a real-world entity. Examples of concepts are gene, protein, author, publication.

Definition 3: P is a set of concept properties. $P = \{p_1, p_2, \ldots \ldots p_n\}$ where a property is an attribute of a concept such as sequence string of a gene or name of a protein.

A property pi $\in$ P is defined as a unary relation of the form p_i (c_i), where $c_i \in$. C is the concept associated with the property pi.

Definition 4: R is a set of semantic relationships between the properties of concepts, where R can be one of the eight types of relationship below. A semantic relation r belongs to R and is defined as a binary relation of the form:

$r = (p_i (c_1), p_j (c_2), RC)$ where $c_1, c_2 \in$ C , and RC is a Relationship closeness value.

A relationship r $(p_i (c_1), p_j (c_2))$ between concept c_1 and c_2 may be one of the following:

i) "is a homolog of" A gene or a protein may share identity with a second gene or protein as they are descended from a common ancestral sequence. The relationship can be calculated using the degree of identity between two sequences.
ii) "is a ortholog of". Orthologs are genes in different species that evolved from a common ancestral gene by speciation. Normally, orthologs retain the same function in the course of evolution. Identification of orthologs is critical for reliable prediction of gene function in newly sequenced genomes. Orthology can be expressed using the homology between either the genes or the proteins they encode.
iii) "is a paralog of". Paralogs are genes related by duplication within a genome. Orthologs retain the same function in the course of evolution, whereas paralogs evolve new functions, even if these are related to the original one. Paralogs can be expressed using the homology between either the genes or the proteins they encode.
iv) "same molecular function". The functions of a gene product are the jobs that it does. A pair of genes could have the same function if they are annotated by an equivalent GO term.
v) "same biological process". A pair of genes could have the same biological process if they are annotated by an equivalent GO term.
vi) "same cellular component". A cellular component describes locations, at the levels of subcellular structures and macromolecular complexes. A pair of genes could have the same cellular component if they are annotated by an equivalent GO term.
Other relationships used in SLM(not defined here due to space limitation) are:"is contained in", "equivalent to","is encoded by", "is transcribed from", "in same family".

Definition 5: G is a set of algorithms. G = {g1,g2,......gn}. These algorithms, such as BLAST, similarity matches and other mining tools, are used to calculate a relationship between concepts. Such algorithms look at all possible pairs of specified concepts from data sources and assign a relationship closeness score to each pair. If this score is above a cut-off or threshold value, then the relationship is accepted. This value can be adjusted for an investigation to increase or decrease the number of matches.

4.2 Formal Representation

A Soft Link Model (SLM) consists of concepts, relationships and degrees of linking. A concept is an entry in a database that represents a real-world entity. The SLM models the linkage between data sources in terms of concepts, properties and semantic relations, and is formally defined as: SLM = (Ci, Cj, R, RC) where Ci, Cj are concepts, R is a type of relationship, and RC is the relationship closeness for the linkage between the concepts. The relationship between two concepts is determined by considering the different properties (pi, pj) of concepts. It can be formed by the syntax: $R = (p_i (c_1), p_j (c_2), g, c)$ where $pj(c_1)$ is a property(for instance, sequence, name) of the first concept, $pj (c_2)$ is a property of the second concept, g is an algorithm used to calculate the relationship and c is a cut-off score or threshold value.

A Soft Link Model (SLM) can be modelled as a graph $G = (V,E)$, where V is a set of nodes and E is a set of edges. Concepts are represented by nodes, relationship types between concepts are represented by edges between nodes, and the closeness is represented by a label under the edge. The label of the node is given by a string which represents a concept name. The label on the edge represents any user defined relationship. Figure 1 shows how the data sources can be connected through the SLM.

Fig. 1. Representation of Soft Link Model

When using the SLM, the following operations are available to the user:

a)Add Relationship to Soft Link model: SLM New = SLM Old U {R}.
b)Remove Relationship to Soft Link model: SLM New = SLM Old - {R}.
c)Add instance to relationship tables. R New = R old U {r}.
d)Remove instance to relationship tables. R New = R old - {r}

A user can suggest a new relationship by providing the System Administrator with the following information: Pair of databases, Pair concepts, Relationship type, Relationship closeness, Pair of identifiers for the DBS(Database Systems).
We implemented in the SLM the following relationships types

• Homolog relationship: Homolog similarity closeness is expressed as the percentage of amino acid sequence identity between the protein sequences of a pair of gene products and is calculated using the BLAST algorithm
• Ortholog relationship: Ortholog relationship closeness is expressed as the percentage amino acid sequence identity between the proteins sequences of a pair of gene products and is calculated using BLAST.
• Related-molecular-function Relationship: Given a pair of gene products, Gi, Gj which are annotated by a set of molecular function terms Ti, Tj respectively, where Ti and Tj consist of m and n terms respectively, the relationship closeness between the genes is calculated as the average of inter-set similarity between terms:

$$RC(Gi, Gj)= (1/m*n)(\sum sim (t_k, t_p).$$

Related-biological-process Relationship and Related-cellular-component Relationship are calculated also as the same as Related-molecular-function Relationship.

4.3   Building of SLM

SLM is built by: 1) Identifying the concepts and properties to be used in the model. 2) Choosing the relationship type between the concepts and properties. 3) Setting up a threshold for the relationship closeness measure. 4) Choosing algorithms to compute the soft link. A successful algorithm is required for the comparison of two concept properties. The variable used to measure the closeness of the biological entity's relationship should also be specified. 5) Creating SLM tables. This can be done in different ways, e.g. offline or on-the-fly at run time.

5   The Integration Framework

Integration of data sources using the SLM model is performed in two phases: Phase 1-relationship discovery and data mining. Phase 2- data source linkage and integration. The system architecture is based on a mediation architecture. Figure 2 and 3 show the functional architecture of our system. We describe briefly its main components:

Phase 1: Relationship discovery System: This phase discovers the relationships between biological objects. These relationships include homology, orthology, paralog, same-function, and others. Many tools are used for discovering such relationships

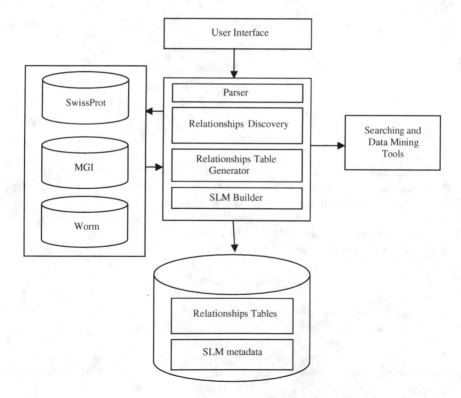

Fig. 2. Architecture of building SLM

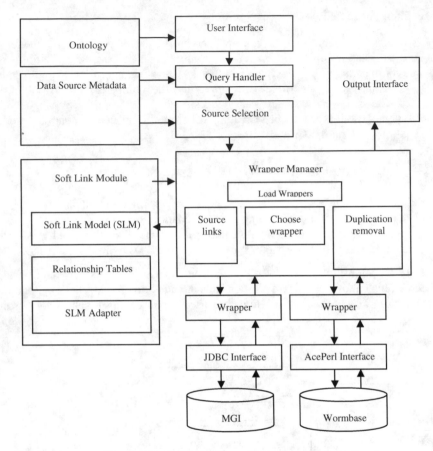

Fig. 3. Overall Architecture of Integration System

eg. Alignment tools, text matching or other data mining tools, such as classification and association rules. The first step is parsing the data sources to extract the attributes of interest which will be used for the relationship. Having identified the concepts and attributes of a source-pair, the system will invoke the appropriate algorithm to discover the existing (if any) relationship between them and these algorithms are responsible for calculating the degree of relationship between the pair. The objects involved in the relationships are then stored in a knowledge base in a triple form (Source id, Target id, Relationship Closeness). User preference (constraints, parameters, algorithms) will be taken into account ·during the discovery of relationships and the linking of building relationship tables. Beside these relationship tables, the relationships metadata will be stored in a relational table as (source, target, relationship type, name of file which contains the actual data).These metadata describe the relationship type, data sources involved and objects involved and refer to the table storing the actual mapping. The user can adjust the parameters used in discovering relationships and calculating degree of relationship.

Phase 2: Data source linkage and Integration: This enables a user to use the relationships to integrate the data. Figure 3 shows the high –level architecture. There are five modules. The Soft Link Module will mine and store the relationships and cross-references between different objects. It provides a flexible linkage between data sources using the discovered relationships in this phase. The source selection module decomposes the query into sub queries that are executed on individual sources and identifies which external sources can answer each sub-query. It will use Relationship tables to link different data sources to answer the user query and then sends the sub-queries to the data sources through wrapper modules. When it receives results from individual sources, it integrates the results and sends them to the user.

The Wrapper manager module is responsible for instantiating the various wrappers the system is configured to use. It manages existing wrappers and performs many tasks: loading existing wrappers, communicating with the Soft link model to retrieve relationships between objects across sources and removing duplication during result assembly from different sources. The wrappers module provides access to remote data sources and transforms the results into an integrated representation. It shields users from the structure and complexity of data sources. There is one wrapper for each data source involved in the system. A number of generalized wrappers have been developed to wrap the heterogeneous data sources. Each wrapper is used to access data of a specific format. If a source allows direct access to its underlying RDBMS, a JDBC wrapper would forward the SQL query statement for processing at the source database. Relationship tables can be used to automatically generate data-base links. The user interface provides a single access point for users to query data-bases within the system. It hides the complexity of the underlying structure and data schema of each database. This interface can accept different types of user queries. The data Source Metadata   module consists of a description part which includes information on how to access data and retrieve it, and a structure part which contains information about logical and physical structure. Each data source's metadata will contain name, URL, description, owner, system, database type and whether there is a direct access to this source or not, and the JDBC driver needed.

6   Case Study

In this section, we provide a worked case study to describe how the relationships can be used to annotate genes and enhance the richness of a dataset in a data analysis process. The data used in the case study comes from gene expression micro-array experiments(Barrett, Suzek et al. 2005). It is benchmark gene expression profile at different stages used to monitor changes in cardiac. We use only the up regulated genes, namely 500 genes.

In this study during the initial process it is essential to determine as much information about the experiment dataset as possible. This information can be detected from column names and their contents as well as other types of metadata held in such datasets. This may require data transformation to a suitable format. The technique used here is described in other author's paper. We create a Soft Link model for the following relationships between mouse and c.elegance: Homolog, Orthology, Molecular function, Biological process, Cellular component. We run our tool for building SLM as described in section 4.3. First we parse all mouse sequence and worm sequences from Swiss-Prot using the parser. Then we use BLAST algorithm to compute the homology between the sequences. For generating Molecular function, biological process and cellular component, we use the algorithm described in (Lord, Stevens et al. 2003). The relationships instances are then stored in a relational table as (source id, target id, relationship closeness). After we have built all relationship tables, we create our SLM as shown in figure 4. Once the SLM has been, the system is ready to use it to annotate gene expression dataset with available information in MGD and Worm data sources. The system will allow the user to select the linkage key of a dataset if he wishes to link this dataset with information across species, namely mouse and c. elegance. Here the system has the flexibility to allow the user to choose the relationships to be used in the linkage and integration. In our case we ran the system several times using different relation-ships, homology relationships next using molecular function, cellular component and biological process with different relationship closeness (RC). The system integrates and links the datasets with data in mouse database. Related genes from other species, namely c. elegance are retrieved based on semantic relationships on SLM. The system processes the query as follow: i) It generates a retrieval query to MGD with the Gene identifier. ii) The wrapper manager extracts the gene identifier from the result set. iii) It generates a new call to the Soft Link Adaptor to retrieve all relationships associated with the Gene concept to other species. iv) Wrapper manager generates a new call to the other species drivers to retrieve all related genes from other sources which have any kind of relationships with the target source. v) It merges retrieved result sets from other species with the result set generated from the target source. vi) It returns the result to the user.

```xml
<?xml version='1.0' encoding='UTF-8'?>
<SLM-knowledge-base no="2">
<database name="mgi">
   <concept Name='Gene_product'>
<relations>
 <SLM DBName='worm' concept='Gene_product'
RelationType='Homolog'  File='homolgy' FileType='mySQL'/>
 </relations>
 <relations>
 <SLM DBName=worm ' concept='Gene_product' RelationType='GO
term(Molecular_function)' File='MF' FileType='mySQL'/>
 </relations>
 <relations>
 <SLM DBName='worm' concept='Gene_product' RelationType='GO
term(Biological process)' File='BP' FileType='mySQL'/>
 </relations>
 <relations>
 <SLM DBName='worm' concept='Gene_product' RelationType='GO
term(cellular_component)' File='cc' FileType='mySQL'/>
 </relations>
 </concept>
</database>.
</SLM-knowledge-base>
```

Fig. 4. An example of SLM metadata

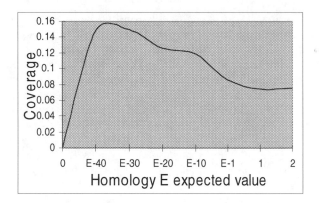

Fig. 5. The correlation between molecular-function relationship and homology relationship.The coverage is number of matches between homology results and molecular function results divided by total of HM results.

Figure 5 shows the correlation between the homology and the molecular function in the result set. The coverage is number of matches between homology results and molecular function results divided by total of HM results. Table 1 show different result set generated by the system using different relationships type and relationship closeness value. These different results allow the biologist to analyse the dataset in different ways. These processes enable the formulation of novel hypotheses leading to informed design of new cycles of laboratory research. When we run our system with same dataset several times with different relationships each time, we got different result set which reflect how the biological objects are connected with each other in different ways. These different results allow biologist to analyse the datasets in different ways and give insight the nature of biological objects. A measure of relationship closeness should afford a biologist a new tool in their repertoire for analysis. The case study has shown how we can use SLM to link a dataset with public data sources in different ways using the relationships to provide data integration within the frame-work of a data analysis process.

Table 1. Result set generated using different relationships

	HM (1E-20)	HM (1E-08)	MF	BP	Overlap (Bp&mf)	Overlap (Bp&HM-08	Overlap (Bp& HM-20)
Result set #	908	1525	25439	24802	825	476	464

Keys: (HM=isHomologyTo, MF = hasSameMolecularFunction, BP = hasSameBiological Process ,CC=hasSameCellularComponent)

7   Discussion

In our survey of biological and bioinformatics literatures, we found the more important relationships between biological objects are: homology, orthology, paralog, and molecular function. These types of relationships are most used for in linking biological objects. We developed a system to link information across species based on these relationships. The novelty of this approach is it allows a user to link a gene expression dataset with heterogeneous data sources dynamically and flexibly.

The idea is to use different relationships with different relationship closeness values to link gene expression datasets with other information available in public bioinformatics data sources. A novel approach based on semantic relationships between bio-logical objects to identify a group of similar genes is presented.

The system is generic so adding new data source is straightforward. Support for new data sources requires the implementation of a data source wrapper module. This wrapper must implement methods used in the general wrapper in the system, and then we need to create the new soft link as described in 4.4.

The key aspects of the SLM approach are the following: i) SLM integrates data from remote sources without bringing the data physically into a central database at the time the researcher needs it. Thus it uses the current version of the data. ii) The biologist can change the linkage type according to their research agenda. Depending on the research questions being asked, the biologist can choose appropriate links between the different concepts and objects. They may experiment with the link type until they get a useful answer for their investigation. Thus the system is flexible and allows different types of links to be used. iii) By using a soft link model (SLM), the system can automatically generate links to other databases that contain further information. iv) SLM provides a method to link genetic databases to other databases, including non-bioinformatics databases, containing information about concepts such as drugs, biochemistry, clinical information and patients. For example, a clinical database may not have a one-to-one mapping with a genetic database, but there is a clear relation-ship which can be presented in this approach. v) SLM allows the user to browse the discovered relationships between data sources. vi) SLM allows the user to browse objects involved in a specific relationship. vii) It stores relationships between sources and exploits them to combine annotation knowledge from different sources.

8 Conclusion

In this paper an approach to the integration of diverse bioinformatics data sources, using a flexible linkage, is presented. Soft Links are modelled via concepts that are interrelated, using a rich set of possible relationship types. The proposed model pursues a novel approach that provides a flexible and soft linkage between data sources. A flexible relationship allows biologist to mine effectively the exponentially increasing amount of comparative genomic information. A preliminary deployment of this technology will be used as a basis to enable cross species functional annotation of data generated by array experiments to inform better the selection of targets for more detailed analysis based on cross species functional information. Furthermore, once the Soft Links are established secondary analysis on genomic elements such as the transcription control elements (transcription factor binding sites) can be analysed to provide novel insights into evolutionary conservation of gene expression. A prototype system based on Java technology for metadata extraction and linking with other public bioinformatics data sources has been implemented; it has been used to identify taxonomically conserved functional pathways. The Soft Link Model (SLM) provides a method to link different databases, including non-bioinformatics databases. In summary we believe the SLM approach allows biologist to access different data sources efficiently using a single system. It also provides a means of linking datasets from other disciplines. Using this approach, a user will not need to be aware of which appropriate data sources to use and how to access them, thus greatly reducing the time and effort taken to analyze their datasets. Future work will regard the full implementation and integration of the SLM, and its extension to different application domains.

References

Aparicio, A. S., O. L. M. Farias, et al. (2005). Applying Ontologies in the Integration of Heterogeneous Relational Databases. Australasian Ontology Workshop (AOW 2005), Sydney, Australia, ACS.

Baxevanis, A. D. and B. F. F. Ouellette, Eds. (2001). Bioinformatics: A Practical Guide to the Analysis of Genes and Proteins. New York, John Wiley & Sons.

Ben-Miled, Z., N. Li, et al. (2004). "On the Integration of a Large Number of Life Science Web Databases." Lecture Notes in Bioinformatics (LNBI): 172-186.

Ben Milad, Z., Y. Liu, et al. (2003). Distributed Databases.

Bleiholder, J., Z. e. Lacroix, et al. (2004). "BioFast: Challenges in Exploring Linked Life Science Sources." SIGMOD Record 33(2): 72-77.

Carel, R. (2003). "Practical Data Integration In Biopharmaceutical Research and Development." PharmaGenomics: 22-35.

Collet, C., M. N. Huhns, et al. (1991). "Resource Integration Using a Large Knowledge Base in Carnot." IEEE Computer 24(12): 55-62.

Davidson, S., J. Crabtree, et al. (2001). "K2/Kleisli and GUS: experiments in integrated access to genomic data sources." IBM Journal

Decker, S., M. Erdmann, et al. (1999). Ontobroker: Ontology Based Access to Distributed and Semi-Structured Information. Database Semantics - Semantic Issues in Multimedia Systems,Proceedings TC2/WG 2.6 8th Working Conference on Database Semantics (DS-8), Rotorua, New Zealand, Kluwer Academic Publishers, Boston.

Dennis, G., Jr., B. T. Sherman, et al. (2003). "DAVID: Database for Annotation, Visualization, and Integrated Discovery." Genome Biol 4(5): P3.

Etzold, T., A. Ulyanov, et al. (1996). "SRS: information retrieval system for molecular biology data banks." Methods Enzymol 266: 114-28.

Freier, A., R. Hofestadt, et al. (2002). "BioDataServer: a SQL-based service for the online integration of life science data." In Silico Biol 2(2): 37-57.

Goble, C., R. Stevens, et al. (2001). "Transparent Access to Multiple Bioinformatics Information Sources." IBM Systems Journal 40(2): 534-551.

Gruber, T. R. (1995). "Toward principles for the design of ontologies used for knowledge sharing." International Journal of HumanComputer Studies 43: 907-928.

Gupta, A., B. Ludäscher, et al. (2000). Knowledge-Based Integration of Neuroscience Data Sources. 12th International Conference on Scientific and Statistical Database Management (SSDBM), Berlin, Germany, IEEE Computer Society.

Heflin, J. and J. Hendler (2000). Dynamic Ontologies on the Web. Proceedings of 17th National Conference on Artificial Intelligence(AAAI-2000). Menlo Park,CA, AAAI/MIT Press.

Kashyap, V. and A. P. Sheth (1996). "Semantic and schematic similarities between database objects: A context-based approach." VLDB Journal: Very Large Data Bases 5(4): 276-304.

Lacroix, Z. and T. Critchlow, Eds. (2003). Bioinformatics: Managing Scientific Data. multimedia information and systems. San Francisco, Morgan Kaufmann.

Leser, U. and F. Naumann (2005). (Almost) Hands-Off Information Integration for the Life Sciences. Proceedings of the Conference in Innovative Database Research (CIDR) 2005, Asilomar, CA.

Necib, C. B. and J. C. Freytag (2004). Using Ontologies for Database Query Reformulation. ADBIS (Local Proceedings).

Rector, A., S. Bechhofer, et al. (1997). "The grail concept modelling language for medical terminology." Artificial Intelligence in Medicine 9: 139-171.

Robert., H. and M. Patricia. (2002). SRS as a possible infrastructure for GBIF. GBIF DADI Meeting. San Diego.

Venkatesh, T. V. and H. Harlow (2002). "Integromics: challenges in data integration." Genome Biology 3(8): reports4027.1 - reports4027.3.

Wache, H., T. V. ogele, et al. (2001). Ontology-Based Integration of Information --- A Survey of Existing Approaches. IJCAI-2001 Workshop on Ontologies and Information Sharing, Seattle, USA.

Wiederhold, G. (1992). "Mediators in the architecture of future information systems." Computer 25(3): 38-49. (2004). "The Genomics Unified Schema(GUS) platform for Functional genomics."

Al-Daihani, B., A. Gray, et al. (2005). Soft Link Model(SLM) for Bioinformatics Data Source Integration. International Symposium on Health Informatics and Bioinformatics, Turkey'05, Antalya, Turkey, Middle East Technical University.

Ashburner, M., C. A. Ball, et al. (2000). "Gene ontology: tool for the unification of biology. The Gene Ontology Consortium." Nat Genet 25(1): 25-9.

Benson, D. A., I. Karsch-Mizrachi, et al. (2005). "GenBank." Nucleic Acids Res 33(Database issue): D34-8.

Bleiholder, J., Z. e. Lacroix, et al. (2004). "BioFast: Challenges in Exploring Linked Life Science Sources." SIGMOD Record 33(2): 72-77.

Buntrock, R. E. (2001). "Chemical registries--in the fourth decade of service." J Chem Inf Comput Sci 41(2): 259-63.

Etzold, T., A. Ulyanov, et al. (1996). "SRS: information retrieval system for molecular biology data banks." Methods Enzymol 266: 114-28.

Freier, A., R. Hofestadt, et al. (2002). "BioDataServer: a SQL-based service for the online integration of life science data." In Silico Biol 2(2): 37-57.

Gupta, A., B. Ludäscher, et al. (2000). Knowledge-Based Integration of Neuroscience Data Sources. 12th International Conference on Scientific and Statistical Database Management (SSDBM), Berlin, Germany, IEEE Computer Society.

Kanz, C., P. Aldebert, et al. (2005). "The EMBL Nucleotide Sequence Database." Nucleic Acids Res 33(Database issue): D29-33.

Kohler, J. (2003). SEMEDA: Ontology based semantic integration of biological data-bases.

Kohler, J. (2004). "Integration of life science databases." BioSlico 2(2): 61-69.

Lacroix, Z. and T. Critchlow, Eds. (2003). Bioinformatics: Managing Scientific Data. multimedia information and systems. San Francisco, Morgan Kaufmann.

Leser, U. and F. Naumann (2005). (Almost) Hands-Off Information Integration for the Life Sciences. Proceedings of the Conference in Innovative Database Re-search (CIDR) 2005, Asilomar, CA.

Letovsky, S. L., Ed. (1999). Bioinformatics: databases and systems, Massachusetts: Kluwer Academic Publishers.

Maglott, D., J. Ostell, et al. (2005). "Entrez Gene: gene-centered information at NCBI." Nucleic Acids Res 33(Database issue): D54-8.

Robbins, R. J. (1995). "Information infrastructure for the human genome pro-ject." IEEE Engineering in Medicine and Biology 14(6): 746--759.

Schneider, M., M. Tognolli, et al. (2004). "The Swiss-Prot protein knowledgebase and ExPASy: providing the plant community with high quality proteomic data and tools." Plant Physiol Biochem 42(12): 1013-21.

Williams, N. (1997). "How to get databases talking the same language." Science 275(5298): 301-2.

Barrett, T., T. O. Suzek, et al. (2005). "NCBI GEO: mining millions of expression profiles--database and tools." Nucl. Acids Res. %R 10.1093/nar/gki022 33(suppl_1): D562-566.

Lord, P. W., R. D. Stevens, et al. (2003). "Investigating semantic similarity measures across the Gene Ontology: the relationship between sequence and annotation." Bioinformatics %R 10.1093/bioinformatics/btg153 19(10): 1275-1283.

An Efficient Storage Model for the SBML Documents Using Object Databases

Seung-Hyun Jung[1], Tae-Sung Jung[1], Tae-Kyung Kim[1], Kyoung-Ran Kim[1],
Jae-Soo Yoo[2], and Wan-Sup Cho[3]

[1] Dept. of Information Industrial Engineering, Chungbuk National University
361-763 Cheongju, Chungbuk, Korea
sane7142@gmail.com,{mispro, tkkim, shira07}@chungbuk.ac.kr
[2] Dept. of Computer and Communication Engineering, Chungbuk National University
361-763 Cheongju, Chungbuk, Korea
yjs@chungbuk.ac.kr
[3] Corresponding author: Dept. of MIS, Chungbuk National University
361-763 Cheongju, Chungbuk, Korea
wscho@chungbuk.ac.kr

Abstract. As SBML is regarded as a de-facto standard to express the biological network data in systems biology, the amount of the SBML documents is exponentially increasing. We propose an SBML data management system (SMS) on top of an object database. Since the object database supports abundant data types like multi-valued attributes and object references, mapping from the SBML documents into the object database is straightforward. We adopt the event-based SAX parser instead of the DOM parser for dealing with huge SBML documents. Note that DOM parser suffers from excessive memory overhead for the document parsing. For high quality data, SMS supports data cleansing function by using gene ontology. Finally, SMS generates user query results in an SBML format (for data exchange) or in a visual graphs (for intuitive understanding). Real experiments show that our approach is superior to the one using conventional relational databases in the aspects of the modeling capability, storage requirements, and data quality.

Keywords: SBML, Object Database, Bioinformatics, Ontology, Pathway, System Biology.

1 Introduction

SBML(Systems Biology Markup Language) is a computer readable XML-based format for representing models of biochemical reaction networks. It is now de-facto standard to express and exchange experimental data in systems biology [8,14]. Lately, SBML documents are widely used in various computational tools that handle cell signaling networks, metabolic pathways, and gene regulatory networks [8,14].

Recently, many databases in systems biology are trying to publish the query results in the SBML format [16] because SBML is regarded as a solution to free up

M.M. Dalkilic, S. Kim, and J. Yang (Eds.): VDMB 2006, LNBI 4316, pp. 94–105, 2006.

biologists from struggling with data formatting issues. As the amount of SBML documents is growing at exponential rate, an efficient management of the SBML documents in a database (rather than files) becomes an important issue in the bioinformatics area.

Conventional simulation or analysis tools use just one SBML document as their input (e.g., cell-designer). However, simulation for the integrated SBML documents from different data sources (e.g., metabolic pathway database and regulatory pathway database) has recently been an important research issue [8,16,7]. SMS provides an efficient integration of the biochemical networks (metabolic, regulatory, or signal transduction pathways) represented by the SBML formats.

We propose an *SBML document Management System (SMS)* based on an object database and gene ontology (GO) [6]. Since SBML itself conforms to the object data model, mapping from SBML documents into an object database is straightforward [5,6]. Furthermore, the object database supports various data types and multi-valued attributes, complex SBML data can be managed easily by the object database compared with a relational database [1]. F. Achard *et al.* [2] recommended that XML and the object database are the best pair to manage and integrate biological data. We verified that the number of classes and the storage space corresponding to the same XML data can be reduced significantly (about 30% and 50% respectively) if we use an object database instead of a relational database [12].

Another important feature is the usage of the SAX parser instead of the DOM parser. Note that the SAX parser requires small memory space compared with DOM parser. In [12], we verified that the memory space required by the SAX parser is about 20% of that of the DOM parser.

The last feature is improved data quality by removing ambiguity and duplication of the terms appeared in the SBML documents. We analyzed 13,000 SBML documents generated by the KEGG2SBML[5] and found that each term (e.g., enzymes, substrates, or products) appears 77 times on the average in the SBML documents. Since this leads to severe data duplication in the database construction, we need a careful sharing of such information instead of the duplication. Furthermore, 30% of the terms in the SBML documents are represented by non-standard terminologies.

To remove duplications and ambiguities of biological terms, *SMS* utilizes GO (Gene Ontology) and data cleansing module (*CM*). During the construction of the object database, 30% of the biochemical terms in the SBML documents are replaced by the representative terms in GO. Moreover, there is significant data duplication in the SBML documents which may bring a severe data inconsistency problem in the database. In SMS, all the data duplication are removed automatically by using excellent database functions (e.g., primary key, unique constraints, and object identifier references).

SMS supports not only the management of the SBML documents in an object database but also graph-based visualization [9] for intuitive understanding and analysis of the SBML documents.

The paper is organized as follows. In Section 2, we present related work. In Section 3, we proposed SBML schema generation in an object database and describe system architecture of SMS. In Section 4, we evaluate SMS in the aspects of storage efficiency and accuracy. In Section 5, we conclude the paper.

2   Related Work

In this section, we present SBML documents, SBML converter which transforms SBML documents into an object database and generates SBML documents from query result for the object database. We then present usage of gene ontology for higher data quality in the database.

2.1   System Biology Markup Language (SBML)

SBML [14] has been proposed by the systems biology community, which consists of major SW makers in systems biology, and becomes a de facto standard language for the representation of biochemical networks. The primary objective of SBML is to represent biochemical reaction networks model, including *cell signaling networks*, *metabolic pathways*, and *gene regulation networks*. SBML provides a standard format for the integration and analysis of biochemical networks among various S/W tools [14].

Each SBML document contains a number of compartments, species, reactions, functions etc. Table 1 shows the main elements of the SBML documents.

Table 1. Elements of the SBML documents

Elements	Description
Compartment	A container of finite volume for well-stirred substances where reactions take place.
Species	A chemical substance or entity that takes part in a reaction. Some example species are ions such as calcium ions and molecules such as ATP.
Reaction	A statement describing some transformation, transport or binding process that can change one or more species. Reactions have associated rate laws describing the manner in which they take place.
Parameter	A quantity that has a symbolic name. SBML provides the ability to define parameters that are global to a model, as well as parameters that are local to a single reaction.
Unit definition	A name for a unit used in the expression of quantities in a model. This is a facility for both setting default units and for allowing combinations of units to be given abbreviated names.
Rule	A mathematical expression that is added to the mode equations constructed from the set of reactions. Rules can be used to set parameter values, establish constraints between quantities, etc.

Fig.1 shows a conceptual biochemical reaction process. The reaction process can be modeled as a network consisting of nodes for compartments or species and links for reactions.

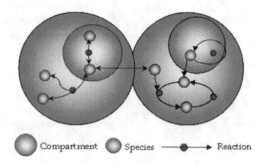

Fig. 1. Conceptual description of an SBML document

2.2 SBML Converter

In systems biology, there are well known databases for the storage and management of the biochemical pathway information: KEGG[10], BioCyc[4], Reactome [13] etc. In KEGG (Kyoto Encyclopedia of Genes and Genomes), a project called KEGG2SBML has been carried out for automatic conversion from pathway database, KGML files, and LIGAND database in KEGG into SBML [5] documents. Fig. 2 shows the structure of KEGG2SBML.

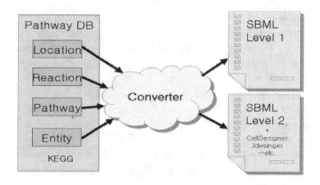

Fig. 2. Structure of KEGG2SBML converter

Various databases or file systems can be used for storing/querying SBML documents: *relational databases, native XML databases, object databases and file systems*. Although a file system provides simple storage, it cannot support sophisticated queries and updates on the SBML documents. A relational database has difficulty in the modeling of complex documents in flat tables, and requires expensive join operations. Note that SBML schema has been defined in an object data model [14]. XML native database may be a good candidate for storing SBML documents, but it is still premature for large SBML documents.

However, object database provides better facilities for storing SBML documents rather than relational database. First, the model similarity between object database and SBML schema reduces the number of the generated tables. Second, object database provides fast query processing as it uses object references instead of join operations.

Third, object database provides good efficiency in storing very large documents as it provides various data types and requires less tables than relational database.

Table 2 shows the advantage of object database for overcoming defect of file systems, relation databases, and native XML databases.

Table 2. Comparison of storage for SBML documents

Storage	Description
Object databases	- Consistent data model between object databases and SBML schema - Simple queries with better query performance
XML databases	- Premature database functionality (query performance, concurrency and recovery etc.) - Lack of useful tools yet (GUI, Development tools etc.)
Relational databases	- Data model mismatch between complex SBML documents and flat tables - Complex queries and query performance overhead
File systems	- Difficulty in elaborate query - Problem of data duplication and security

2.3 Gene Ontology (GO)

Ontology is a knowledge model including vocabulary, concept, relationships etc. of a specific field. In other words, it is a specification of a conceptualization of knowledge in the domain [15,11]. It defines standard terms and creates relationships among the information.

The aim of GO is to service controlled vocabularies for the description of the molecular functions, biological processes, and cellular components of gene products. These terms are to be utilized such as attributes of gene products by collaborating databases, facilitating uniform queries across them. GO is a hierarchical classification scheme structured as a direct acyclic graph (DAG) [3].

Fig.3. shows GO in a graph form. Here, the table *term* defines terminologies, the table *term_synonym* stores synonyms of the terms, and the table *term2term* stores relationships among the terms.

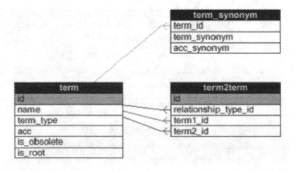

Fig. 3. Scheme of gene ontology database

3 SMS: SBML Management System

In this section, we will describe system architecture for SMS and schema mapping in detail.

3.1 SBML Schema Mapping

To store SBML documents in an object database, the first step required is schema mapping from SBML schema into an object schema. Since SBML schema has been defined as an object oriented data model, the mapping is straightforward.

We have many advantages when an object database is used for the storage of large SBML documents. Table 3 shows UML notation of SBML schema. Note that there 8 multi-valued attributes whose domains are another class. Table 4 shows corresponding object database schema implemented by UniSQL [17] object database management system. Since UniSQL directly supports multi-valued attributes, implementation of the SBML schema is straightforward. All information in Table 3 can be stored in a class (Table 4) with 8 multi-valued attributes. In the case of relational databases, they require 8 additional tables (one for each multi-valued attribute) for the SBML schema in Table 3; furthermore, the queries for the tables becomes very complex because of the joining for the 8 tables. This also imposes performance overhead especially for the large SBML database.

Table 3. UML notation of SBML scheme

Model

id : Oid {use="optional"}
name : string {use="optional"}
functionDefinition : FunctionDefinition [0...*]
unitDefinition : UnitDefinition [0...*]
compartment : Compartment [0...*]
species : Species [0...*]
parameter : Parameter [0...*]
rule : Rule [0...*]
reaction : Reaction [0...*]
event : Event [0...*]

Table 4. Object-Oriented scheme

Model

id : Oid
name : String
functionDefinition : FunctionDefinition {SET}
unitDefinition : UnitDefinition {SET}
compartment : Compartment {SET}
species : Species {SET}
parameter : Parameter {SET}
rule : Rule {SET}
reaction : Reaction {SET}
event : Event {SET}

3.2 SMS Architecture and Main Features

As shown in Fig. 4, *SMS* has four components: *Local Databases (Pathway Database and GO Database), SBML Converter, Data Cleansing Module (CM),* and *Result Visualization Module (VM). Local Databases* have been constructed from public biochemical pathway databases and GO databases. Data extraction from those databases has been accomplished by an *SBML converter* (when the database publishes SBML documents) or a mediator (when the database publishes its own data format). During data extraction from various biochemical network databases, each terminology can be replaced by a standard terminology based on the GO database. Terminology used in the query can also be replaced by a standard or synonymous terminology if needed. *SBML Converter* generates query results either in SBML documents or in a graph format. In the SBML converter, we used the SAX parser instead of the DOM parser because of the memory efficiency. As we verified in the literature [12], memory space required by the DOM parser is three times larger than that of the SAX parser.

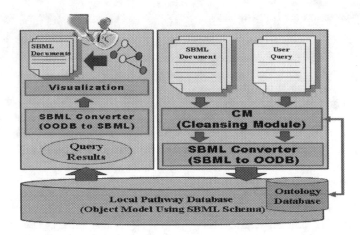

Fig. 4. The Structure of SMS

3.3 Improvement of Data Quality by Utilizing Data Cleaning Module (CM)

For higher data quality, we developed a data cleaning module *CM* in *SMS*. Since we construct a local pathway database from various public source pathway databases, there may be a serious *data redundancy* and *non-standard terminologies* problems in the database.

We analyzed 13,000 SBML documents generated by the KEGG2SBML[5] and found that each term (e.g., enzymes, substrates, or products) appears 70 times (data redundancy) on the average. This may lead to severe data duplication in the database construction. In SMS, data cleaning module solves this problem by removing duplication. All the data duplication can be replaced automatically by object references. Excellent database functions such as primary key, unique constraints, and object identifier references can be used. During the construction of the object database, 30% of the biochemical terms in the SBML documents are replaced by the representative terms in GO.

Fig. 5 shows the data cleaning module in detail. Let assume that the elements *A* and *A'*, which are synonyms, appear two documents *D1* and *D2*. The data cleaning module receives the document *D1* and stores the element *A* into the database. Then it receives the document *D2* and replaces *A'* with the representative terminology *A*. Then before storing *A* into the local database, data cleaning module checks whether *A* is already in the local database or not. Since *A* is already in the database, duplication occurs and data cleaning module discards the insertion of *A* into the database. For the document B, reference to the object A is used instead of new object generation.

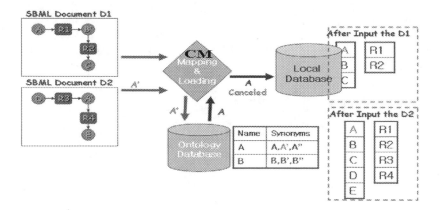

Fig. 5. To remove duplication terms, *SMS* adopted *CM* and *GO*

3.4   Examples

SMS has been used in two ways in the metabolic pathway reconstruction system [9] as shown in Fig. 6. The first one is the import of SBML documents regardless of their source databases. SBML documents from various metabolic pathway databases (KEGG, EcoCyc, etc.) are analyzed by SMS and the extracted data can be stored in the local database. Note that as the number of pathway databases providing their data in SBML format increases, SMS will be more useful. For non-SBML documents, we need various mediators to convert the pathway data into the local database. The second one is the export of the query result in SBML format to the users or other useful tools. Since there are many tools in systems biology (more than 90 SW packages) which accept SBML as their input data format, generation of query result in SBML format is very useful for data exchange among them.

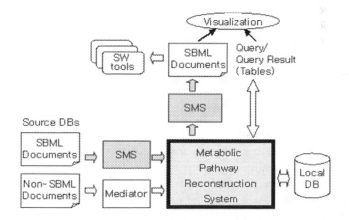

Fig. 6. Metabolic pathway reconstruction system using SMS

4 Evaluation

In this section, we evaluate *SMS* compared with conventional systems. We then present how the data quality can be improved in the *SMS* by utilizing *CM*.

4.1 Comparison with Conventional Systems

Table 5 shows comparison *SMS* with KEGG2SBML, an SBML conversion tools by KEGG[10]. KEGG2SBML changes the metabolic pathway data in KEGG into SBML format. It is developed only for converting from KEGG data to SBML document files for purpose of the data exchange.

SMS is a unique SBML management system to overcome the functional limitation of KEGG2SBML. Compared with KEGG2SBML, *SMS* stores huge SBML documents into an object database and publishes the query results in an SBML format. Furthermore, *SMS* guarantees high quality of the pathway database by removing duplicated data with gene ontology.

Table 5. Comparison to KEGG2SBML and SMS

	KEGG2SBML	SMS
Importing SBML	X	O
Exporting SBML	O	O
GUI	X	O
SBML data management	X	O
Validation of biochemical terms	X	O
Platform	Linux	Windows, Linux, Unix

4.2 Evaluation of the Data Quality

SMS improved data quality by removing ambiguity and duplication of the terms appeared in the SBML documents. We analyzed 12,220 SBML documents generated by the KEGG2SBML[5] and found that each biochemical term (e.g., enzymes, substrates, or products) appears 77 times on the average in the SBML documents. Since this leads to severe data duplication in the database construction, we need a careful sharing of such information instead of the duplication. Furthermore, 30% of the terms in the SBML documents are represented by non-standard terminologies.

To remove duplications and ambiguities of biological terms, *SMS* utilizes GO (Gene Ontology) and data cleansing module (*CM*). During the construction of the object database, 30% of the biochemical terms in the SBML documents are replaced by the representative terms in GO. Moreover, there is significant data duplication in the SBML documents which may bring a severe data inconsistency problem in the database. In SMS, all the data duplication has been removed automatically by using excellent database functions (e.g., primary key, unique constraints, and object references).

Table 6 shows data duplications and non-standard terminologies in the SBML documents generated by KEGG2SBML.

Table 6. Result of data quality improvement via CM

	Total	Duplication degree	Non-standard terminologies
Biochemical Terms	2,897	Average 77	750 terms
Reactions	1,862	Average 50	x

4.3 Comparison of Storage Systems - Object Database and Relational Database

A real experiment shows that an object database has many advantages compared with relational databases. We constructed both object database and relational database for the same SBML documents from KEGG. Fig. 7 (a) and (b) show the comparison of the relational schema (11 tables) and object schema (4 classes).

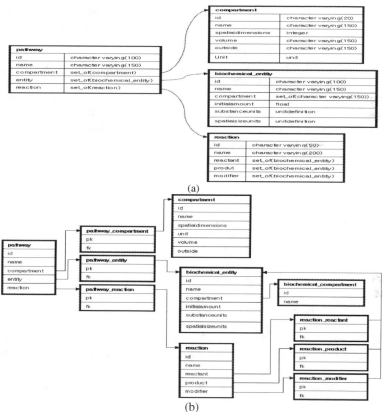

(a) Object database schema. (b) Relational database schema.

Fig. 7. Comparison of database schema

Table 7 shows summary of the comparisons. The number of tables and records are reduced significantly (40%) in the object database. The object database requires 4 classes as it utilized the object references (OIDs) and multi-valued attributes. However, the relational database has 11 tables because it requires additional relations to normalize multi-valued attributes. Moreover, the number of the tables affects the number of the records and actual storage size. Note that the object database requires less than half of the storage size of the relational database.

Table 7. Space Requirement Comparison

Database Data Size	Object Database			Relational Database			The Rate Of Decrease (%) = ⓐ / ⓑ
	Class	ⓐObject	Storage Size (MB)	Table	ⓑRecord	Storage Size(MB)	
80 MB	4	104,037	52M	11	229,064	92M	45.4 %
150 MB	4	208,074	106M	11	520,128	213M	41.6%
240 MB	4	321,111	150M	11	857,192	322M	38.4%

5 Conclusion

We proposed an SBML document management system called *SMS* for efficient biochemical data management on top of an object database. We adequately utilized the object references (OID) and multi-valued attributes for simple storing of SBML data into the object database. Note that model similarity between object database and SBML documents reduces the complexity of the database in the aspect of the number of the tables and the size of the storage. For efficient storing of huge SBML documents into the database, we adopt event-based SAX parse instead of the DOM parser for minimal memory utilization. We utilized gene ontology in the system to remove terminology ambiguity. Query result can be published in a visual graph by using GUI tools and represented in an SBML format.

Acknowledgement. This work was supported by the Regional Research Centers Program of the Ministry of Education & Human Resources Development in Korea.

References

[1] K. Aberer, "The use of object-oriented data models in biomolecular databases," In Proc. International Conference on Object-Oriented Computing in the Natural Sciences, Heidelberg, Cermany, 1994

[2] F. Achard, et al., "XML, bioinformatics and data integration," Bioinformatics Review, 2001

[3] G. Battista, et al., "Drawing Directed Acyclic Graphs: An Experimental Study," *International Journal of Computational Geometry and Applications 10,* 2000. 6.

[4] Ecocyc Web Site, http://ecocyc.org/

[5] A. Funahashi and H. Kitano, "Converting KEGG DB to SBML," Bioinformatics, 2003. 6.

[6] Gene Ontology, http://www.geneontology.org

[7] M. Hucka, et al., "The systems Biology Markup Language (SBML): a medium for representation and exchange of biochemical network models," Systems Biology Workbench Development Group, 2002. 10.

[8]   M. Hucka, et al., "Systems Biology Markup Language (SBML) Level 2 : Structrues and Facilities for Model Definitions," Systems Biology Workbench Development Group, 2003. 6

[9]   T. S. Jung, et al., "A Unified Object Database for Various Biochemical Pathways," Bioinfo2005, 2005

[10]   KEGG Web Site, http://www.genome.jp/kegg/

[11]   J. Kohler, et al., "SEMEDA: ontology based semantic integration of biological databases," Bioinformatics, 2003. 12.

[12]   T. K. Kim and W. S. Cho, "A DTD-dependent XML Data Management System : An Object-Relational Approach," In Proc. International Conference on Internet & Multimedia Systems & Applications (IMSA 2005), Hawaii, 248-253

[13]   Reactome Web Site, http://www.genomeknowledge.org/

[14]   SBML Web Site, http://www.sbml.org.

[15]   E. Shoop, et al., "Cosimini Data exploration tools for the gene ontology database," Bioinformatics, 2004. 7.

[16]   L. Stromba and P. Lambrix, "Representations of molecular pathways: an evaluation of SBML, PSI MI and BioPAX," Bioinfotmatics, 2005

[17]   UniSQL Web Site, http://www.cubrid.com

Identification of Phenotype-Defining Gene Signatures Using the Gene-Pair Matrix Based Clustering

Chung-Wein Lee, Shuyu Dan Li, Eric W. Su, and Birong Liao*

Integrative Biology, Lilly Research Laboratories, Lilly Corporate Center,
Indianapolis, IN 46285
Liao_Birong@lilly.com

Abstract. Mining the "meaningful" clues from vast amount of expression profiling data remains to be challenge for biologists. After all the statistical tests, biologists often struggle deciding how to do next with a large list of genes without any obvious theme of mechanism, partly because most statistical analyses do not incorporate understanding of biological systems before hand. Here, we developed a novel method of "gene –pair difference within a sample" to identify phenotype-defining gene signatures, based on the hypothesis that a biological state is governed by the relative difference among different biological processes. For gene expression, it is relative difference among the genes within a sample (an individual, cell, etc), the highest frequency of occurrences a gene contributing to the within sample difference underline the contributions of genes in defining the biological states. We tested the method on three datasets, and identified the most important gene-pairs to drive the phenotypic differences.

Keyword: Gene Pair, hierarchical clustering, phenotype-defining gene signatures, lymphomas and adenocarcinoma.

1 Introduction

Whole genome expression profiling has been broadly applied in basic and applied biological research, especially in the area of cancer subtype classification, predication of survival after chemotherapy, and design of diagnostic signature for patient response [1-4]. Recently, obtaining expression profiles has become relatively easy, as demonstrated by the increasing amount of data deposited in the public repositories [5]. However, understanding of biological mechanism underlining the profile and the interpretation of vast amount of data have been constant challenges to biologists [6].

Ever since hierarchical clustering was introduced into organizing the expression profile data by Eisen [7], it has become the method of choice for the attempt to understand biological mechanisms of the expression profile in a given genetics or environmental background. Despite many flavors of algorithm and statistical validation methods have been developed[8], most biologists are still relying heavily

---

* Corresponding author.

M.M. Dalkilic, S. Kim, and J. Yang (Eds.): VDMB 2006, LNBI 4316, pp. 106 – 119, 2006.
© Springer-Verlag Berlin Heidelberg 2006

on visual inspection and their prior biological knowledge to decipher the "meaning" of clustering results. It has been recommended that biologists should use multiple methods to reveal underlining data structure and exploitation of biological mechanisms [9-11]. After all the statistical or mathematical manipulations of data, biologists are often left with a large list of genes without any obvious theme of mechanism. To decide the next action to follow up the result has been a very subjective task.

Biological systems are robust, with many redundancies built in [12-15]. The redundancy may be part of reason that very little overlap between list of statistically significant genes from two studies on the same biological system [16]. Recently, using the idea of biologically defined set of genes to compare two studies on lung cancer, Subramanian et al (2005) were able to identify the commonality of mechanism of poor outcome between the studies.

Almost all of initial microarray analyses, either two-color or single-color, are first to normalize the expression data to a fixed value. In the case of Affymetrix platform, the number is called global scaling (www.affymetrix.com) [17], while many other two-color systems set to zero the summation of logarithmical ratios of one channel to another (Cy3 or Cy5 dye) [18]. Typically, one would identify genes of the statistical significance by comparing the fluorescent value of a gene across the experimental conditions. A gene was chosen if it met a statistical threshold of difference between the conditions. We were hypothesizing that a biological state was governed by the relative difference among different biological processes within an individual (cell, patient, etc.). In the case of expression, it will be relative difference among the genes within a sample (rather than across samples), the highest frequency of occurrences a gene contributing to the within sample difference may have underlining biological meaning. Therefore, the compilation of relative difference across all the biological samples would help biologists to prioritize the list of genes and would also help to interpret the biological mechanisms of difference across the conditions. In this paper, we used a novel method of "gene –pair difference within a sample" to illustrate the point, using three different data sets. We were able to identify the most important gene-pairs to drive the phenotypic differences.

2 Materials and Methods

2.1 Data Sets

The expression data were downloaded from accompanying web sites of supplemental information of the respective publications (Table 1).

2.2 Data Processing and Analysis

For expression data on Affymetrix platform, all .CEL files were downloaded and reanalyzed in Affymetrix MAS 5 software, with a global scale of 500. All other analyses, including hierarchical clustering with average linkage and Kaplan-Meier survival estimates were carried out in Matlab (The MathWorks, Inc). Algorithms are omitted here due to the space. Interested readers are referred to the manual.

Table 1. Gene expression profiling data analyzed in this study

Reference	Cancer type	Array platform	Sample size	URL for data downloading
Rosenwald et al (2002)	DLBCL	cDNA (Lymphochip)	240	http://llmpp.nih.gov/
Monti et al (2005)	DLBCL	Affymetrix (U133A, U133B)	176	http://www.broad.mit. edu/cgi- bin/cancer/datasets.cgi
Bhattacharjee et al (2001)	Lung Adenocar cinoma	Affymetrix (U95Av2)	139	http://www.broad.mit. edu/cgi- bin/cancer/datasets.cgi

To explore the mechanism based clustering with "gene-pair differences within a sample", we first identified the data sets with the obvious phenotypes. In the area of cancer research, the "gold" standard for subtyping is the clinical outcome after various drug treatments. An average microarray contains thousands of genes, to explore the all possible gene pairs will be an astronomically large number that an average desktop computer will not be able to handle the computation. For example, Affymetrix U133A and B contains more that 44K probe sets, they will result in 1 billion pairs. In addition, almost all experiments will have a substantial number of genes whose expressions are not detected, which is evidenced by the "absent" call in Affymetrix platform or "zero or missing value" in two-color systems. The process is following:

- Filter the non-reliable intensity value across all samples (1/3 of "absent" or missing values, etc)
- Calculate differences among all gene pairs with a sample (for log transformed data, the arithmetic difference was used, whereas intensity data were log transformed first before the arithmetic difference calculation)
- Rank all possible pair differences with the highest standard deviation across all samples
- Reiterate clustering the samples using pair differences until optimum separation of phenotype achieved.
- Survival analyses among the cluster to find the optimal contrast
- Obtain the number of gene pairs for the optimal contrast
- Count the frequency of participating genes

3 Results

The method was tested on three datasets. These datasets were chosen because they contain substantial amount of phenotypic information (subtype, patient survival, etc).

Diffuse Large B-cell Lymphomas (Rosenwald data set). The dataset contains expression profile of 7399 spots (most clones have duplicate spots) on 240 patients [19]. The data are in log2 (Cy5/Cy3) with Cy5 representing the experimental samples and Cy3 representing reference pool. Hierarchical clustering was performed on the 100 to 300K gene pairs with highest standard deviation. The dendrograms of 5K to 100 K showed consistent tree structures (data not showed), therefore, 5K pairs were chosen for further analysis. Figure 1 showed the dendrogram of highest 5K pairs, which had a STD cutoff of ~3. It can be clearly divided into three distinct clusters as suggested by the original data generators. Survival analysis with the patient clinical outcome data among these three clusters showed distinct profiles (Figure 2). In contrast, the dendrogram of pairs with low STD values had no consistent structure (data not showed).

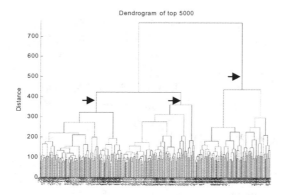

Fig. 1. Hierarchical clustering on the differences among gene pairs clearly revealed the three distinct clusters of 240 patients. Arrow indicates the cluster separating point.

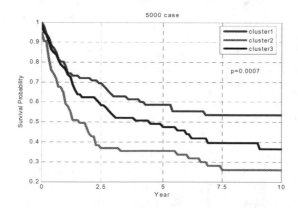

Fig. 2. Kaplan-Meier Survival analyses demonstrated distinct survival profiles with p value of 0.007 for the clusters from Figure 1

We next sought to examine the gene compositions of 5K gene pairs. There were total 3667 probe sets (as defined by different UNIQID number) representing ~ 50% of total probes in the microarray, since there were multiple spots for each gene, they accounted for 1989 genes. The dendrogram based upon expression of 3667 probe sets did not show clear grouping as that of 5K gene pairs (data not shown). When the frequency of probe sets participating in 5K gene pairs was examined, we found that large difference among them. Most of them only participated in 5K gene pair once. Table 2 showed the frequency of top 50 probe sets. As obvious from the table, the top ranking genes showed consistency in their contribution to 5K gene pairs as showed by high frequency of different spot representing the same genes. Interestingly, Ig mu participated in ~4000 pairs out of 5K pairs, suggesting the importance of this molecule in the disease.

The next question was if these frequently participated genes also showed the great pair difference in their expressions, thus contributing to separation of sub groups. We examined the ranking from the biggest variation in the pairs across the samples, Table 3 showed the 5 highest gene pairs. Not surprisingly, Ig mu participated in all 5 pairs, whereas rhombotin-like 1 gene participated three times.

Two-hundred forty patients were then equally divided into three groups according to expression value of Ig mu. Kaplan-Meier analysis was performed on these groups. Figure 3 showed that the contrast among survival curve of these three groups were not statistically significant ($p = 0.25$). However, when the value of pairs between Ig mu and others were used, the significance of the contrast among the groups were statistically improved, an example was showed in Figure 4.

Diffuse Large B-cell Lymphomas (Monti et al data set). The data contains expression profile of ~44K probe sets (~33K gene, Affymetrix U133A and B arrays) on 176 patients[20]. 7429 out of 47 million gene pairs with standard deviation across all patient samples greater than 3.0 were selected. The dendrograms based on 500 to all 7429 gene pairs showed clear separation among the different clusters (Figure 5). We then used Kaplan-Meier survival analysis with available patient survival data and the clusters generated from the expression differential from 100 gene pairs to 7429 pairs to find the best contrast among the clusters. Although the survival contrasts were generally good for all size of pairs, top 500 pairs showed the best contrast (Figure 6). The composition of 500 pairs was further analyzed. RGS13 was found to participate in the pairs for 287 times, whereas Ig Mu ranked the third and contributed to 24 pairs.

There were at least two ways of DLBCL subtype classifications based upon the gene expressions. One is the cell-of-origin analysis developed by Rosewald et al (2002), where three subtypes were identified, i. e. activate b-cell-like subtype (ABC), germinal-center b-cell-like (GCB) and type III. Another is the gene set enrichment analysis (GSEA) used in Monti et al. (2005), where three subtypes were suggested to be "oxidative phosphorylation (OxPhos)", "B-cell receptor/proliferation (BCR)", and "host-response (HR)". Table 4 illustrated the relationship between clusters from gene pair method and the above mentioned two methods, and suggested cluster 1 of gene pair method with high percentage of GCB and BCR and cluster 2 with high percentage of Type III and HR.

Table 2. Frequency of individual probe that contribute to separation of clusters

UID	Description	Frequency
19326	immunoglobulin heavy constant mu	3646
33225	immunoglobulin heavy constant mu	236
34404	immunoglobulin kappa constant	205
16516	T-cell leukemia/lymphoma 1A	165
17723	immunoglobulin kappa constant	143
24291	T-cell leukemia/lymphoma 1A	108
28826	T-cell leukemia/lymphoma 1A	96
27032	immunoglobulin kappa constant	82
19333	immunoglobulin kappa constant	72
34314	immunoglobulin kappa constant	61
27840	immunoglobulin kappa constant	60
24376	ESTs, Weakly similar to A47224 thyroxine-binding globulin precursor [H.sapiens]	57
33200	immunoglobulin heavy constant mu	39
16049	immunoglobulin heavy constant mu	36
17239	immunoglobulin kappa constant	29
27157	cytochrome P450, subfamily XXVIIA (steroid 27-hydroxylase, cerebrotendinous xanthomatosis), polypeptide 1	17
30727	LC_30727	17
27294	matrix metalloproteinase 7 (matrilysin, uterine)	16
23952	fibronectin 1	16
25029	LC_25029	15
28979	LIM domain only 2 (rhombotin-like 1)	15
19238	LIM domain only 2 (rhombotin-like 1)	15
24953	Homo sapiens cDNA: FLJ22747 fis, clone KAIA0120	14
17331	Fc fragment of IgG, high affinity Ia, receptor for (CD64)	14
29312	secreted phosphoprotein 1 (osteopontin, bone sialoprotein I, early T-lymphocyte activation 1)	14
17218	LIM domain only 2 (rhombotin-like 1)	14
29222	LC_29222	14
25237	complement component (3d/Epstein Barr virus) receptor 2	13
34350	Homo sapiens, clone MGC:24130 IMAGE:4692359, mRNA, complete cds	13
19379	fibronectin 1	13
28489	small inducible cytokine subfamily A (Cys-Cys), member 18, pulmonary and activation-regulated	12
27413	Fc fragment of IgG, high affinity Ia, receptor for (CD64)	12
19370	Homo sapiens, clone MGC:24130 IMAGE:4692359, mRNA, complete cds	12

Table 2. *(Continued)*

17354	signaling lymphocytic activation molecule	12
17791	fibronectin 1	12
19361	collagen, type III, alpha 1 (Ehlers-Danlos syndrome type IV, autosomal dominant)	12
28655	monokine induced by gamma interferon	11
32626	monokine induced by gamma interferon	11
34269	heat shock 70kD protein 1A	11
17475	heat shock 70kD protein 1A	11
26361	Homo sapiens, clone MGC:24130 IMAGE:4692359, mRNA, complete cds	11
27093	cytochrome P450, subfamily XXVIIA (steroid 27-hydroxylase, cerebrotendinous xanthomatosis), polypeptide 1	11
19384	mitogen-activated protein kinase 10	11
26118	fructose-1,6-bisphosphatase 1	11
17364	complement component (3d/Epstein Barr virus) receptor 2	10
30724	pre-B lymphocyte gene 3	10
17496	v-myb myeloblastosis viral oncogene homolog (avian)-like 1	10
27499	KIAA0233 gene product	10
16016	fibronectin 1	10

Table 3. The statistics of expression values of five highest gene pair difference across 240 patient samples

Pair Rank	UID	Accession	Description	Highest	Lowest	Average	Stdev	Range
	19326	Hs.153261	immunoglobulin heavy constant mu	4.468	-6.982	-0.67	2.69	11.45
1	24376	Hs.317970	ESTs, Weakly similar to A47224 thyroxine-binding globulin precursor [H.sapiens]	6.117	-3.447	0.69	2.22	9.564
2	28979	Hs.184585	LIM domain only 2 (rhombotin-like 1)	3.908	-4.832	-0.16	1.68	8.74
3	27294	Hs.2256	matrix metalloproteinase 7 (matrilysin, uterine)	8.37	-2.741	0.54	2.08	11.111
4	19238	Hs.184585	LIM domain only 2 (rhombotin-like 1)	2.923	-4.069	-0.05	1.62	6.992
5	17218	Hs.184585	LIM domain only 2 (rhombotin-like 1)	2.353	-4.23	-0.2	1.64	6.583

Note: all 5 highest gene pairs involve immunoglobulin heavy constant mu

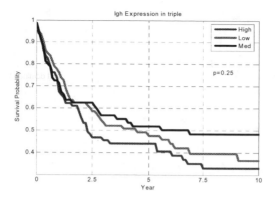

Fig. 3. Expression of Ig mu alone statistically did not separate groups well (p=0.25). Two-hundred forty patients were equally divided into three groups according to expression value of IgH mu. Kaplan-Meier analysis was performed on these groups.

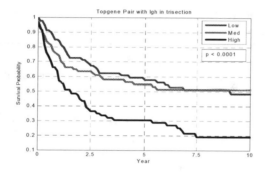

Fig. 4. Kaplan-Meier analysis of gene pair between Ig mu and the gene that is weakly similar to A47224 thyroxine-binding globulin precursor

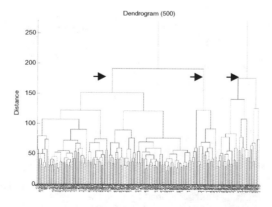

Fig. 5. Hierarchical clustering on the differences among gene pairs revealed the three distinct clusters of 176 patients. Arrows indicates the cluster separating point.

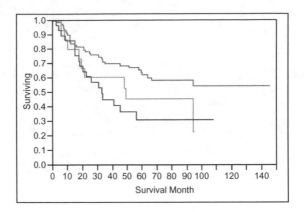

Fig. 6. Kaplan-Meier Survival analyses demonstrated distinct survival profiles for the clusters from Figure 5, with a Log-Rank p value of 0.02

Lung adenocarcinoma (Bhattacharjee data set). The data set was derived from [21], which contains expression profile of 190 lung adenocarcinoma patients. However, only 113 patients come with clinical data, thus only their expression profile (159 microarrays, some patient samples have multiple runs) was subject to gene pair analysis. Hierarchical clustering was performed on the 50 gene pairs with stable separation of phenotype (Figure 7). Survival analyses of these clusters were performed to find the best contrast (Figure 8). We found that the contrast among different clusters were generally robust even with a relatively small number of gene pairs (Figure 8), with a log-rank p value of 0.0848. Two genes were identified to participate in the separation with high frequency, namely, fibrinogen gamma chain and pulmonary-associated protein C.

Table 4. The relationship of clusters from 500 pairs and clusters of other DLBCL classifications

Method	Cluster 1	Cluster 2	Cluster 3
Cell-of-origin*			
GCB	66	8	11
ABC	25	5	4
TypeIII	31	7	19
GSEA**			
OxPhos	37	8	5
HR	26	5	18
BCR	59	7	11

* Cell-Of-Origin method developed by Rosenwald et al (2002).
** Gene set enrichment analysis used in Monti et al. (2005).

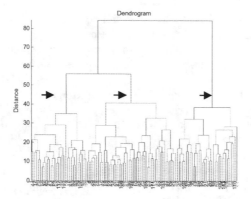

Fig. 7. Hierarchical clustering on the expression differentials among 50 gene pairs revealed the three distinct subclass of adenocarcinoma. Arrow indicates the cluster separating points.

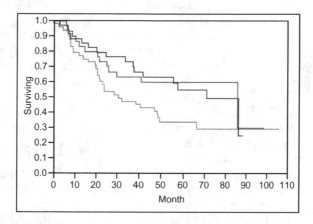

Fig. 8. Kaplan-Meier Survival analyses demonstrated survival profiles for the clusters from Figure 7, with a Log-Rank p value of 0.08

4 Discussion

Due to the fact that almost all microarray analyses, either two-color or single-color, normalize initially the expression data within a given hybridization, either through global scaling in the case of Affymetrix technology (www.affymetrix.com) [22], or through zeroing the sum of ratios one fluorescent channel to the others in two-color systems [23], the assumption that the fluorescent intensity represent some true expression value (for example, the number of copy for a specific transcript) is compromised. This is especially challenging when one wants to compare the value across biological samples under various treatments. Hypothetically, it is very likely that the sample intensity value may represent very different copy numbers of a transcript, simply because the copy numbers of other transcripts in a biological sample have changed. To alleviate this challenge, we devised a method of "gene pair difference within a sample" to identify the phenotype-defining signatures, based upon the hypothesis that a biological state was governed by the relative difference among different biological processes within an individual (cell, patient, etc.).   Using the three public datasets, we were able to identify gene-pairs and genes that drive the phenotypic differences.

In Rosenwald's dataset [24], clustering on the differences of 5000 gene pairs gives clear grouping (Figure 1), furthermore, these 5000 pairs are contributed from about 50% total probes in the microarray. Kaplan-Meier analyses using the survival data from patients, in combination of grouping information from the pair difference demonstrate that the difference in phenotype (in this case, survival) are always better separated by the pairs rather than the original gene expression value (Figures 2, 3). When these phenotype-defining pairs are further examined, an interesting phenomenon revealed, immunoglobulin heavy constant mu(IghM) has the highest appearance among all genes from 5000 pairs (Table 2), indicating the importance of IghM in defining DLBCL patient survival.   This is reinforced by the observation that multiple clones representing IghM in "lymphochip" have higher appearance than other genes (Table 2). IghM is one of 100 markers that were used to classify different subtypes of DLBCL patients [25,26]. Indeed, IghM is important for B-cell development as mutations in IghM have been implicated in B-cell malignancy [27,28].   Another gene that frequently participates in the gene pairs with IghM to drive the phenotype separation is rhomtotin-like 1, it's altered expression has also been documented to play an important role in b-cell development and T-cell malignancy [29].

When the method was applied to another DLBCL dataset [30], three genes were identified to frequently participate in the phenotype-defining signature. They are regulator of G-protein signaling 13 (RGS13), IghM and another (229152_at) with no annotation so far. Although the exact role of RGS13 in DLBCL is not yet clear, however, it has recently been shown to differentially express in IgM-stimulated B-cell lymphomas vs germinal center B cell lymphomas [31], and in mantle cell lymphoma ([32].   Interestingly, germinal center B cell lymphomas is one of three subtypes organized by Rosewald et al (2002) using gene-expression patterns. As Table 4 shows, there were considerable agreement among the gene-pair methods, cell-of-origin method [33] and gene set enrichment analysis method[34,35]. Cluster 1 of gene pair method is highly concentrated with samples from GCB and BCR, and cluster 2 with samples from Type III and HR.

In the lung dataset again, the subtypes can be robustly classified even with small number of gene pairs (Figures 7 and 8). Two genes that frequently participate in the subtype-defining gene pair signatures are identified, namely, fibrinogen gamma chain and surfactant pulmonary-associated protein C (SFTPC). Surfactant protein C is exclusively expressed by alveolar type II epithelial cells of lung, mutations in the coding gene have been linked to chronic lung disease both in adults and children [36]. Little is known about the role of fibrinogen gamma chain in lung cancer, however, the elevated levels of this protein have been found in the plasma of lung cancer patients[37].

In the practical biomarker discovery, one of important aspect is to find a small number of molecules so they can be tracked with the established molecular techniques. Combining phenotypes and method of "gene-pair difference within a sample", we are able to find the relatively small number of molecules that participate in the relatively large amount of phenotype-defining gene pair signature. There are some limitations of this method, it is not for new class discovery as many researches using global expression profiling to find a new subtype. Rather, the method is used for identification of significant contributors for the well-defined phenotypes. Like all other statistical and mathematic methods, the incompleteness of probes in a microarray, quality of probe, measuring sensitivity, etc. will impact the identification of phenotype-defining gene pair signatures. For example, RGS13 is one of most important contributors for the subtype classification in Monti dataset, not in Rosewald dataset, since the two dataset using different platforms of microarray. When examining the expression value of this gene in two dataset, we found that the values was more robust in Monti dataset (data not shown), possibly due to probe performance. However, this method offers the advantage of combining basic biological understanding of technology and statistical analyses, much improvement can be made on this concept and more research in this area is highly needed.

Acknowledgement

We thank anonymous reviewers for their helpful comments.

References

1. Savage KJ, Monti S, Kutok JL, Cattoretti G, Neuberg D, De Leval L et al.: The molecular signature of mediastinal large B-cell lymphoma differs from that of other diffuse large B-cell lymphomas and shares features with classical Hodgkin lymphoma. Blood 2003, 102: 3871-3879.
2. Rosenwald A, Wright G, Chan WC, Connors JM, Campo E, Fisher RI et al.: The use of molecular profiling to predict survival after chemotherapy for diffuse large-B-cell lymphoma. N Engl J Med 2002, %20;346: 1937-1947.
3. Alizadeh AA, Eisen MB, Davis RE, Ma C, Lossos IS, Rosenwald A et al.: Distinct types of diffuse large B-cell lymphoma identified by gene expression profiling. Nature 2000, 403: 503-511.
4. Lossos IS, Czerwinski DK, Alizadeh AA, Wechser MA, Tibshirani R, Botstein D et al.: Prediction of survival in diffuse large-B-cell lymphoma based on the expression of six genes. N Engl J Med 2004, 350: 1828-1837.
5. Barrett T, Suzek TO, Troup DB, Wilhite SE, Ngau WC, Ledoux P et al.: NCBI GEO: mining millions of expression profiles--database and tools. Nucleic Acids Res 2005, 33 Database Issue: D562-D566.
6. Subramanian A, Tamayo P, Mootha VK, Mukherjee S, Ebert BL, Gillette MA et al.: Gene set enrichment analysis: A knowledge-based approach for interpreting genome-wide expression profiles. PNAS 2005, 0506580102.
7. Eisen MB, Spellman PT, Brown PO, Botstein D: Cluster analysis and display of genome-wide expression patterns. Proc Natl Acad Sci U S A 1998, 95: 14863-14868.
8. Handl J, Knowles J, Kell DB: Computational cluster validation in post-genomic data analysis. Bioinformatics 2005, 21: 3201-3212.

9. Handl J, Knowles J, Kell DB: Computational cluster validation in post-genomic data analysis. Bioinformatics 2005, 21: 3201-3212.

10. Man MZ, Dyson G, Johnson K, Liao B: Evaluating methods for classifying expression data. J Biopharm Stat 2004, 14: 1065-1084.

11. Subramanian A, Tamayo P, Mootha VK, Mukherjee S, Ebert BL, Gillette MA et al.: Gene set enrichment analysis: A knowledge-based approach for interpreting genome-wide expression profiles. PNAS 2005, 0506580102.

12. Oltvai ZN, Barabasi AL: Systems biology. Life's complexity pyramid. Science 2002, 298: 763-764.

13. Ravasz E, Somera AL, Mongru DA, Oltvai ZN, Barabasi AL: Hierarchical organization of modularity in metabolic networks. Science 2002, 297: 1551-1555.

14. Dezso Z, Oltvai ZN, Barabasi AL: Bioinformatics analysis of experimentally determined protein complexes in the yeast Saccharomyces cerevisiae. Genome Res 2003, 13: 2450-2454.

15. Wuchty S, Oltvai ZN, Barabasi AL: Evolutionary conservation of motif constituents in the yeast protein interaction network. Nat Genet 2003, 35: 176-179.

16. Fortunel NO, Otu HH, Ng HH, Chen J, Mu X, Chevassut T et al.: Comment on " 'Stemness': transcriptional profiling of embryonic and adult stem cells" and "a stem cell molecular signature". Science 2003, 302: 393.

17. Liu WM, Mei R, Di X, Ryder TB, Hubbell E, Dee S et al.: Analysis of high density expression microarrays with signed-rank call algorithms. Bioinformatics 2002, 18: 1593-1599.

18. Schena M, Shalon D, Davis RW, Brown PO: Quantitative monitoring of gene expression patterns with a complementary DNA microarray. Science 1995, %20;270: 467-470.

19. Rosenwald A, Wright G, Chan WC, Connors JM, Campo E, Fisher RI et al.: The use of molecular profiling to predict survival after chemotherapy for diffuse large-B-cell lymphoma. N Engl J Med 2002, %20;346: 1937-1947.

20. Monti S, Savage KJ, Kutok JL, Feuerhake F, Kurtin P, Mihm M et al.: Molecular profiling of diffuse large B-cell lymphoma identifies robust subtypes including one characterized by host inflammatory response. Blood 2005, 105: 1851-1861.

21. Bhattacharjee A, Richards WG, Staunton J, Li C, Monti S, Vasa P et al.: Classification of human lung carcinomas by mRNA expression profiling reveals distinct adenocarcinoma subclasses. Proc Natl Acad Sci U S A 2001, 98: 13790-13795.

22. Liu WM, Mei R, Di X, Ryder TB, Hubbell E, Dee S et al.: Analysis of high density expression microarrays with signed-rank call algorithms. Bioinformatics 2002, 18: 1593-1599.

23. Schena M, Shalon D, Davis RW, Brown PO: Quantitative monitoring of gene expression patterns with a complementary DNA microarray. Science 1995, %20;270: 467-470.

24. Rosenwald A, Wright G, Chan WC, Connors JM, Campo E, Fisher RI et al.: The use of molecular profiling to predict survival after chemotherapy for diffuse large-B-cell lymphoma. N Engl J Med 2002, %20;346: 1937-1947.

25. Alizadeh AA, Eisen MB, Davis RE, Ma C, Lossos IS, Rosenwald A et al.: Distinct types of diffuse large B-cell lymphoma identified by gene expression profiling. Nature 2000, 403: 503-511.

26. Rosenwald A, Wright G, Chan WC, Connors JM, Campo E, Fisher RI et al.: The use of molecular profiling to predict survival after chemotherapy for diffuse large-B-cell lymphoma. N Engl J Med 2002, %20;346: 1937-1947.

27. Milili M, Antunes H, Blanco-Betancourt C, Nogueiras A, Santos E, Vasconcelos J et al.: A new case of autosomal recessive agammaglobulinaemia with impaired pre-B cell differentiation due to a large deletion of the IGH locus. European Journal of Pediatrics 2002, 161: 479-484.

28. Lopez GE, Porpiglia AS, Hogan MB, Matamoros N, Krasovec S, Pignata C et al.: Clinical and molecular analysis of patients with defects in micro heavy chain gene. Journal of Clinical Investigation 2002, 110: 1029-1035.

29. Foroni L, Boehm T, White L, Forster A, Sherrington P, Liao XB et al.: The rhombotin gene family encode related LIM-domain proteins whose differing expression suggests multiple roles in mouse development. J Mol Biol 1992, 226: 747-761.

30. Monti S, Savage KJ, Kutok JL, Feuerhake F, Kurtin P, Mihm M et al.: Molecular profiling of diffuse large B-cell lymphoma identifies robust subtypes including one characterized by host inflammatory response. Blood 2005, 105: 1851-1861.

31. Cahir-McFarland ED, Carter K, Rosenwald A, Giltnane JM, Henrickson SE, Staudt LM et al.: Role of NF-kappa B in cell survival and transcription of latent membrane protein 1-expressing or Epstein-Barr virus latency III-infected cells. Journal of Virology 2004, 78: 4108-4119.

32. Islam TC, Asplund AC, Lindvall JM, Nygren L, Liden J, Kimby E et al.: High level of cannabinoid receptor 1, absence of regulator of G protein signalling 13 and differential expression of Cyclin D1 in mantle cell lymphoma. Leukemia 2003, 17: 1880-1890.

33. Rosenwald A, Wright G, Chan WC, Connors JM, Campo E, Fisher RI et al.: The use of molecular profiling to predict survival after chemotherapy for diffuse large-B-cell lymphoma. N Engl J Med 2002, %20;346: 1937-1947.

34. Subramanian A, Tamayo P, Mootha VK, Mukherjee S, Ebert BL, Gillette MA et al.: Gene set enrichment analysis: A knowledge-based approach for interpreting genome-wide expression profiles. PNAS 2005, 0506580102.

35. Monti S, Savage KJ, Kutok JL, Feuerhake F, Kurtin P, Mihm M et al.: Molecular profiling of diffuse large B-cell lymphoma identifies robust subtypes including one characterized by host inflammatory response. Blood 2005, 105: 1851-1861.

36. Bridges JP, Wert SE, Nogee LM, Weaver TE: Expression of a human surfactant protein C mutation associated with interstitial lung disease disrupts lung development in transgenic mice. Journal of Biological Chemistry 2003, 278: 52739-52746.

37. Vejda S, Posovszky C, Zelzer S, Peter B, Bayer E, Gelbmann D et al.: Plasma from cancer patients featuring a characteristic protein composition mediates protection against apoptosis. Mol Cell Proteomics 2002, 1: 387-393.

TP+Close: Mining Frequent Closed Patterns in Gene Expression Datasets

YuQing Miao[1,2], GuoLiang Chen[1], Bin Song[3],
and ZhiHao Wang[1]

[1] Department of Computer Science and Technology, University of Science and Technology
of China, Hefei, China
{myq, glchen, wangzhh}@ustc.edu.cn
[2] Department of Computer Science and Technology, Guilin University of Electronic
Technology, Guilin, China
[3] Department of Computer Science, Case Western Reserve University, Cleveland, USA
bin.song@case.edu

Abstract. Unlike the traditional datasets, gene expression datasets typically contain a huge number of items and few transactions. Though there were a large number of algorithms that had been developed for mining frequent closed patterns, their running time increased exponentially with the average length of the transactions increasing. Therefore, most current methods for high-dimensional gene expression datasets were impractical. In this paper, we proposed a new data structure, tidset-prefix-plus tree (TP+-tree), to store the compressed transposed table of dataset. Based on TP+-tree, an algorithm, TP+close, was developed for mining frequent closed patterns in gene expression datasets. TP+close adopted top-down and divide-and-conquer search strategies on the transaction space. Moreover, TP+close combined efficient pruning and effective optimizing methods. Several experiments on real-life gene expression datasets showed that TP+close was faster than RERII and CARPENTER, two existing algorithms.

Keywords: data mining, association rules, frequent closed pattern, gene expression data, top-down.

1 Introduction

Frequent closed pattern mining plays an essential role in mining association rules from gene expression datasets and discovering biclustering of gene expression [1, 2]. Unlike the traditional datasets, gene expression datasets typically contain a huge number of items (genes) and few transactions (samples). Though a large number of algorithms had been developed for frequent closed pattern mining [3, 4, 5, 6], their running time increased exponentially with the average length of the transactions increasing. Therefore, most current methods for high-dimensional gene expression datasets were impractical [8, 9].

Several mining frequent closed pattern algorithms were specially designed to handle gene expression datasets [7, 8, 9]. The authors of [7] introduced the notion of free sets. They transposed the original dataset and mined the transposed matrix to find the

M.M. Dalkilic, S. Kim, and J. Yang (Eds.): VDMB 2006, LNBI 4316, pp. 120–130, 2006.

free sets on samples. After computing the closures of the free sets, they got the closed sets on genes by Galois connection. Paper [8] proposed the row enumeration tree, and developed an algorithm, CARPENTER, to explore the row enumeration space by constructing projected transposed tables recursively. RERII [9] was another algorithm on the row enumeration tree, it utilized set intersection operations on horizontal layout data. Both CARPENTER and RERII used bottom-up search strategy to explore the transaction space. When the datasets were dense, the methods of CARPENTER and RERII were time-consuming.

In this paper, we will develop a new efficient algorithm, TP+close, for mining frequent closed pattern in gene expression datasets. Algorithm TP+close is inspired by FP-tree [4] and algorithm CHARM [6]. TP+close uses a tidset-prefix-plus tree (TP+-tree) structure to store compressed transposed table of dataset. TP+close adopts top-down and divide-and-conquer search strategies on the transaction space and uses IT-tree (itemset-tidset tree) to explore both the itemset space and transaction space simultaneously. TP+close combines efficient pruning and effective optimizing techniques. The result of experiments shows that the TP+close is much faster than RERII and CARPENTER by up to one order of magnitude.

2 Preliminaries

2.1 Problem Statement

Let $I = \{i_1, i_2, ..., i_m\}$ be a set of items, and D be a transaction dataset, where each transaction has a unique identifier (tid) and contains a subset of I. The set of all tids is denoted as T. A set $X \subseteq I$ is called an itemset, and a set $Y \subseteq T$ is called a tidset. For convenience we write an itemset $\{aeh\}$ as aeh, and a tidset $\{24\}$ as 24. For an itemset X, we denote its corresponding tidset as $t(X)$, i.e., the set of all tids that contain X as a subset. For a tidset Y, we denote its corresponding itemset as $i(Y)$, i.e., the set of items common to all the tids in Y. Note that $t(X)=\cap_{x \in X} t(x)$, $i(Y)=\cap_{y \in Y} i(y)$, and for any two itemsets X and Y, if $X \subseteq Y$, then $t(X) \supseteq t(Y)$ [6].

The support of an itemset X, denoted as $\sigma(X)$, is the number of transaction in which X occurs as a subset. An itemset is frequent if its support is no less than a user-specified minimum support threshold ($min\sigma$), i.e., if $\sigma(X) \geq min\sigma$. A frequent itemset(or a frequent pattern) X is called closed if there exists no proper superset $Y \supset X$ with $\sigma(X)=\sigma(Y)$[6].

By above definition, we will know that $\sigma(X)=|t(X)|$, and itemset X is closed if and only if $X=i(t(X))$, let $c(X)=i \circ t(X)=i(t(X))$, i.e., iff $X=c(X)$.

Example 1: Fig. 1 shows a dataset and its transposed table (TT). There are five transactions and fifteen different items in the dataset (Fig. 1(a)). Table in Fig. 1(b) is a transposed version of table in Fig. 1(a), i.e., the corresponding tidsets of all items. For example, $t(aeh)=t(a) \cap t(e) \cap t(h)=1234 \cap 234 \cap 234=234$, $c(aeh)=i(t(aeh))=i(234)=i(2) \cap i(3) \cap i(4)=adehlpr \cap acehoqt \cap aehpr=aeh$, thus aeh is a closed itemset. $\sigma(aeh)=|t(aeh)|=|234|=3$, if $min\sigma=2$, then aeh is a frequent closed pattern.

Given a gene expression dataset D, which contains n samples (transactions) that are the subset of a set of m genes (items), where $n<<m$, our problem is to discover all frequent closed patterns with respect to a user-specified minimum support threshold ($min\sigma$).

tid$_k$	X_k
1	a,b,c,l,o,s
2	a,d,e,h,l,p,r
3	a,c,e,h,o,q,t
4	a,e,h,p,r
5	b,d,f,g,l,q,s,t

(a)

i$_j$	$t(i_j)$
a	1,2,3,4
b	1,5
c	1,3
d	2,5
e	2,3,4
f	5
g	5
h	2,3,4
l	1,2,5
o	1,3
p	2,4
q	3,5
r	2,4
s	1,5
t	3,5

(b)

Fig. 1. A Dataset D (a) and a Transposed Table TT (b)

2.2 IT-Tree

Let I be the set of items. Define a function $p(X,k)=X[1:k]$ as the k length prefix of X, and a prefix-based equivalence relation θ on itemsets as follows: $\forall X,Y\subseteq I$, $X\theta Y\Leftrightarrow p(X,k)=p(Y,k)$. That is, two itemsets are in the same class if they share a common k length prefix [6].

CHARM [6] searched for frequent closed patterns over an IT-tree (itemset-tidset tree) search space (Fig. 2 shows an example). Each node in the IT-tree, represented by an itemset-tidset pair (IT-pair), $X\times t(X)$, is a prefix-based class. All the children of a given node X belong to its equivalence class. An equivalence class is denoted as $[P]=\{l_1, l_2,\ldots, l_n\}$, where P is the parent node (the prefix), and each l_i is a single item, representing the child node $Pl_i\times t(Pl_i)$.

For any two nodes in the IT-tree, $X_i\times t(X_i)$ and $X_j\times t(X_j)$, if $X_i\subseteq X_j$ then $t(X_i)\supseteq t(X_j)$. Assume that we are currently processing a node $P\times t(P)$ where $[P]=\{l_1, l_2,\ldots, l_n\}$ is the prefix class. Let X_i denote the itemset Pl_i, then each member of $[P]$ is an IT-pair $X_i\times t(X_i)$.

Lemma 1[6]. Let $X_i\times t(X_i)$ and $X_j\times t(X_j)$ be any two member of a class $[P]$, with $X_i\leq_f X_j$, where f is a total order. The following four properties hold:

(1) If $t(X_i) = t(X_j)$, then $c(X_i) = c(X_j) = c(X_i\cup X_j)$
(2) If $t(X_i)\subset t(X_j)$, then $c(X_i)\neq c(X_j)$, but $c(X_i)= c(X_i\cup X_j)$
(3) If $t(X_i)\supset t(X_j)$, then $c(X_i)\neq c(X_j)$, but $c(X_j)= c(X_i\cup X_j)$
(4) If $t(X_i)\neq t(X_j)$, then $c(X_i)\neq c(X_j)\neq c(X_i\cup X_j)$

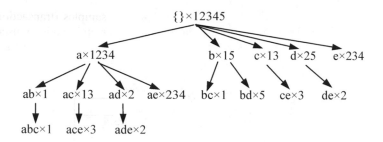

Fig. 2. IT-tree: Itemset-Tidset Search Tree

3 TP+Close Algorithm Design and Implementation

3.1 TP+-Tree and Class

TP+close uses a tidset-prefix-plus tree (TP+-tree) to store compressed transposed table of dataset. TP+-tree is a prefix tree that stores the tids. The corresponding tid-sets of multiple items share a common tid prefix, according to some sorted order of tids. Obviously, if the tids are sorted in their frequency descending order, more prefix strings can be shared. To facilitate tree traversal, a tid header table is built in which each *tid* points to its occurrence in the tree via a *head of side-link*. Nodes with the same *tid* are linked in sequence via such *side-links*.

Definition 1. A tidset-prefix-plus tree (TP+-tree) is a tree structure defined below.

1. It consists of one root labeled as *"null"*, a set of tid prefix subtrees as the children of the root, and a tid header table.
2. Each node in the tid prefix subtree consists of four fields: *tid, level, side-link*, and *item-link*, where *tid* stores which tid this node represents; *level* stores which level this node is in the TP+-tree, the level of root is the zero, the level of the children of the root is the first, follow as such; *side-link* links to the next node in the TP+-tree carrying the same *tid*, or null if there is none; *item-link* links a items list in which these items are just stored, in whose tidsets the *tid* of this node is the last one according to the frequency descending order of tids, or null if there is none.
3. Each entry in the tid header table consists of two fields: *tid* and *head*, which points to the first node in the TP+-tree carrying the *tid*, nodes in the TP+-tree with the same *tid* are linked in increasing order of their level via *side-link*.

Fig. 3 shows a TP+-tree constructed from Fig. 1(b) *TT*. The number in tree node indicates the *tid*, the number in brackets after *tid* indicates the *level*, the lower-case letters after brackets indicate the items list linked by *item-link*, and the line of dashes indicates the *side-link*.

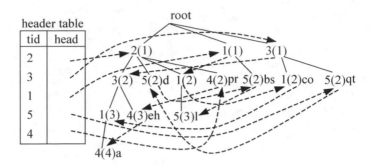

Fig. 3. TP+-tree

The pseudo code of TP+-tree construction algorithm is given in algorithm 1.

Algorithm 1. Const_TP+-tree (*TT*, *min*σ)

Input: The gene expression dataset (i.e., the transposed table *TT* of dataset *D*) and a user-specified minimum support threshold *min*σ.

Output: TP+-tree of *D*

Step:

1.Scan the gene expression dataset *TT* once. Collect the set of frequent items and their support. At the same time, collect the number of frequent items in each transaction (i.e., the cardinality of transaction);

2.Sort the set of tid in descending order of cardinality of transaction as *L* and store it in header table;

3.Create the root of a TP+-tree, *T*, and label it as "*null*". For each frequent item in *TT* do the following:

Sort the tidset of item according to the order of *L*. Let the sorted tidset of item be *uU*, where *u* is the first element and *U* is the remaining tidset, *H* indicates the current position in header table. Call insert_tree(*uU*,*T*).

```
procedure insert_tree(uU,T)
begin
(1)  if T has a children node N such that N.tid=u then{
(2)     if u is the last tid in tidset of the item then
insert this item into N.item-link
     }
(3)  else{
(4)     create a new node N, let N.tid=u,
N.level=T.level+1, N.parent=T
(5)     insert N into the side-link whose header pointer
is H.head , where H.tid=u
(6)     if u is the last tid in tidset of the item then
insert this item into N.item-link
 (7)     else let N.item-link=null
     }
(8)  if U≠∅ then insert_tree(U,N)
end.
```

From the TP+-tree construction process, we can see that exactly two scans of the gene expression dataset are needed: the first scan collects the set of frequent items and the number of frequent items in each transaction, and the second constructs the TP+-tree. Furthermore, two important properties of TP+-tree can be observed.

Lemma 2. Given a gene expression dataset *TT* and a used-specified minimum support threshold *minσ*, the TP+-tree contains the complete information of frequent items of *TT* and their corresponding tidsets, and each item is stored in TP+-tree only once.

Lemma 3. Given a gene expression dataset *TT*, without considering the root, the size of a TP+-tree is bounded by the overall occurrences of the tids in all frequent items of *TT*, and the height of the tree is bounded by the maximal number of tids in any frequent item of *TT*.

Lemma 2 and lemma 3 clarify that TP+-tree is a highly compact structure that stores the complete information of transposed table of dataset.

In TP+-tree, from root to each node along the path, we could get an order tidset. For example, from root to the node "4(4)*a*", we get the order tidset "2314". Let *Q* be the set of all order tidsets found in TP+-tree as such, the tid class is defined as follows.

Define a function $s(U,k)$ as the *k* length suffix of *U*, and a suffix-based equivalence relation *β* on *Q* as follows: $\forall U,V \in Q$, $U\beta V \Leftrightarrow s(U,1)=s(V,1)$. That is, two tidsets are in the same class if they have identical the last tid. We denote an equivalence class as $[S]=\{s_1,s_2,\dots, s_n\}$, where *S* is the tid (the suffix), and each s_i is a tidset in which *S* is the last tid. For example, by searching the TP+-tree in Fig. 3, we can find five tid classes: $[2]=\{2\}$, $[3]=\{3,23\}$, $[1]=\{1,21,31,231\}$, $[5]=\{35,25,15,215\}$, $[4]=\{24,234,2314\}$.

3.2 Algorithm TP+Close

Based on lemma 2, we know that TP+-tree stores all frequent items and their tidsets information, how the itemset of each node is got? Because each item is stored in a leaf or a node just once, that is to say the complete itemsets of all leaves can be got straight. In order to get the itemsets of all nodes, we take a method that searches the entry in descending order of entry's rank in the header table. First, start from the last entry in the header table; search all nodes whose *tid* are same as the entry's along the *side-link*. Because these nodes are leaves, the itemsets of these nodes are got easily. After getting the itemsets of these leaves from their *item-link,* the *item-link* of these leaves are linked to the *item-link* of their parents. Similarly, search all nodes whose *tid* are same as the next entry's, as soon as getting the itemset of a node, the *item-link* of this node will be linked to the *item-link* of its parent. By doing so, we could get the itemsets of all nodes in the TP+-tree.

For any node in the TP+-tree, we could get an itemset and a tidset (from root to the node along path), we use the notation $X\times q(X)$, an itemset-tidset pair, to represent a member of a tid class. Note that $q(X)$ is not same as $t(X)$, $q(X)$ is the prefix of $t(X)$, i.e., $q(X)=p(t(X),k)$, where $1 \leq k \leq |t(X)|$. So we call such an itemset-tidset a quasi-IT-pair.

Once getting the quasi-IT-pair of each member of a tid class, we can perform a depth-first search over the IT-tree of the tid class for frequent closed patterns. After search the IT-trees of all tid classes, all frequent closed patterns will be discovered. Based on the above idea, the pseudo code of TP+close algorithm is given in algorithm 2.

Algorithm 2. TP+close $(TT, min\sigma)$

Input: The transposed table TT of a dataset D and a user-specified minimum support threshold $min\sigma$

Output: All frequent closed patterns and their support, FCP

Step:

1. Call Const_TP+-tree($TT, min\sigma$) for construct the TP+-tree of D;

2. Search a entry in descending order of its rank in header table, start from the last entry in header table;

3. Find all nodes whose tid are same as the entry's in the header table and whose $level \geqslant min\sigma$ in the TP+-tree, get these nodes' quasi-IT-pair, these quasi-IT-pairs form the set N; //pruning 2

4. Insert the found nodes' $item-link$ to their parents' $item-link$, delete the found nodes from TP+-tree;

5. Call SubTP+close(N);

6. Search the next entry, if the entry's rank in header table< $min\sigma$ then stop, else goto step 3. //pruning 1

procedure SubTP+close (N)
begin

(1)   for each node $n_i \times q(n_i)$ in N // n_i as the itemset, $q(n_i)$ as the tidset

(2)   for each node $n_j \times q(n_j)$ in N, with $n_i <. n_j$ // $<.$ is the order of node generated in N

(3) $n' = n_i \cup n_j$, $q(n') = q(n_i) \cap q(n_j)$

(4)   if $|q(n')| \geqslant min\sigma$ then   // pruning 3

(5)   if $q(n_i) = q(n_j)$ then // property 1

(6)   remove $n_j \times q(n_j)$ from N, replace all n_i with n'

(7)   if $q(n_i) \subset q(n_j)$ then // property 2

(8)   replace all n_i with n'

(9)   if $q(n_i) \supset q(n_j)$ and $q(n')$ is not discovered then // property 3 and redundant pruning 1

(10)   add $n' \times q(n')$ to N', remove $n_j \times q(n_j)$ from N // N' stores the subnodes of N

(11)   if $q(n_i) \neq q(n_j)$ and $q(n')$ is not discovered then // property 4 and redundant pruning 1

(12)   add $n' \times q(n')$ to N'

(13)   endif

(14)   endfor

```
(15)    if N'≠∅ then SubTP+close(N')
(16)    if nᵢ is not discovered and not subsumed then //
redundant pruning 2
(17)       add nᵢ× q(nᵢ) to FCP// FCP stores all frequent
closed patterns and their tidset
(18) endfor
end.
```

In algorithm TP+close, support pruning consists of three stages:

Pruning 1. If the rank of an entry in header tables less than $min\sigma$, it is not necessary to search those nodes in TP+-tree whose *tid* are same as the entry's.

Pruning 2. If a node's *level* in TP+-tree less than $min\sigma$, there is no need to get the quasi-IT-pair of this node.

Pruning 3. Give a node $n \times q(n)$ of IT-tree, for a child node $n' \times q(n')$ of $n \times q(n)$, if $|q(n')| < min\sigma$, there is no need to enumerate any more below node $n' \times q(n')$.

Redundant pruning 1. Give a node $n' \times q(n')$ of IT-tree, if $q(n')$ has already been discovered in an earlier enumeration, let it be $n \times q(n)$, with $q(n) = q(n')$. Since $n \supset n'$, there is no need to enumerate any more below node $n' \times q(n')$.

Redundant pruning 2. Since we generate the tid class in descending order of the rank of tid in header table. When we want to add $X \times q(X)$ to *FCP*, it probably happens that $Y \times q(Y)$ has been in *FCP*, with $q(X) \subset q(Y)$ and with $X \subseteq Y$. In this case, X is a redundant set or a non-closed set subsumed by Y, thus it should not be added to *FCP*.

Since TP+close performs union operation on itemsets and intersection operation on tidsets in the IT-tree, the cost of operation on itemsets become inordinately large, and the huge size of itemsets cannot be afforded when the cardinality of itemset gets larger. Therefore we utilize a horizontal bitmap to store itemset, where a bit represents an item.

Theorem 1. TP+close enumerates all frequent closed patterns.

Proof: Since all frequent items and their tidsets are stored in TP+-tree, according to the tid class generation process, for any frequent item X, $X \times t(X)$ and $X \times p(t(X),k)$ are contained in different tid classes based on $t(X)$ or $p(t(X),k)$, where $1 \leq k \leq |t(X)| -1$. After search the IT-trees of all tid classes, by lemma 1, we know that TP+close enumerates all the closed patterns and some non-closed patterns. By support pruning, those patterns that do not have sufficient support are pruned. Further, by redundant pruning, TP+close eliminates any redundant patterns and non-closed patterns subsumed by certain closed patterns. Thus TP+close correctly discovers all the frequent closed patterns.

4   Experiment Evaluation

In this section, we compare the performance of TP+close against RERII and CARPENTER. All our experiments were performed on a PC with a Pentium Ⅲ 1.4 Ghz CPU, 1GB RAM running Windows XP. Algorithms were coded in C++.

Our experiments are performed on 3 real-life datasets, which are the ALL_GCT [10], ALL_AML [10] and DLBCL [11]. In the ALL_GCT dataset, there are 73 tissue samples and each sample is described by the activity level of 255 genes. In the ALL_AML dataset, there are 38 tissue samples and each sample is described by the activity level of 5000 genes. In the DLBCL dataset, there are 176 tissue samples and each sample is described by the activity level of 3000 genes. The DLBCL dataset is discretized by the value P/A/M (present/absent/don't know). The ALL dataset is discretized by a ratio-average method. First, the first column gene is treated as a contrast sample, and the log ratio of each gene expression value is computed. Second, comparing each gene expression value with the average of all value in one sample, if the value is twice of the average, we say the gene is highly expressed, if the value is two times less than the average, we say the gene is highly repressed. In this case, each gene is denoted as two label, one represents the highly expressed of gene, another represents the highly repressed of gene.

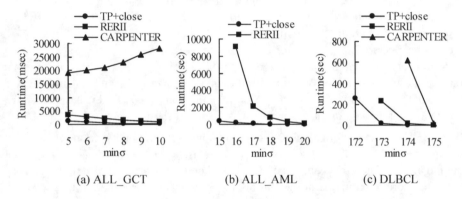

(a) ALL_GCT   (b) ALL_AML   (c) DLBCL

Fig. 4. Runtimes of Algorithms

Fig. 4 shows the experimental results on three datasets. At some points in Fig. 4(b) and 4(c), the runtime of CARPENTER and RERII are not shown because they are too slow above 2 hours and cannot be finished. Among the three algorithms, we find TP+close is generally faster than RERII, while RERII is usually much faster than CARPENTER. In Fig. 4(b) and 4(c), TP+close is one order of magnitude faster than RERII at low support. Table 1, Table 2 and Table 3 list the number of frequent closed patterns are discovered in three datasets. From the three tables, we can see that ALL_GCT dataset is sparse; ALL_AML and DLBCL datasets are dense.

Table 1. The number of frequent closed patterns in ALL_GCT

Minimum support(%)	5(7%)	6(8%)	7(10%)	8(11%)	9(12%)	10(14%)
the number of frequent closed patterns	3525	2464	1657	1074	697	468

Table 2. The number of frequent closed patterns in ALL_AML

Minimum support(%)	15(41%)	16(43%)	17(46%)	18(49%)	19(51%)	20(54%)
the number of frequent closed patterns	18772	8954	4308	2153	1091	548

Table 3. The number of frequent closed patterns in DLBCL

Minimum support(%)	172(97.7%)	173(98.3%)	174(98.9%)	175(99.4%)
the number of frequent closed patterns	15179	2085	242	21

5 Conclusions

In this paper, we have proposed a new data structure, tidset-prefix-plus tree (TP+-tree), for storing compressed transposed table of dataset, and developed an algorithm, TP+close, for mining frequent closed patterns in gene expression datasets. TP+close adopted top-down and divide-and-conquer search techniques on the transaction space and used IT-tree to explore both the itemset space and transaction space. Several experiments showed that TP+close was faster than RERII and CARPENTER by up to one order of magnitude.

There are a lot of interesting research issues related to mining frequent closed patterns in gene expression datasets, including further study of highly scalable algorithms, finding more interesting rules that biologists are interested in, combining advantages of top-down and bottom-up search strategies, and so on.

Acknowledgments. This work was supported in part by the National Natural Science Foundation of China under Grant No.60533020. We thank Mehmet M. Dalkilic, Sun Kim and Jiong Yang for their most helpful advice and comments.

References

1. Creighton, C., Hanash, S.: Mining Gene Expression Databases for Association Rules. Bioinformatics, Vol.19. (2003) 79–86
2. Madeira, S., Oliveira, A.: Biclustering Algorithm for Biological Data Analysis: A Survey. IEEE/ACM Transactions on Computational Biology and Bioinformatics, Vol, 1. (2004) 24–45
3. Agrawal, R., Srikant, R.: Fast Algorithms for Mining Association Rules. Proc. 1994 VLDB Int'l Conf. Santiago, Chile. (1994) 487–499
4. Han, J.W., Pei, J., Yin Y.: Mining Frequent Patterns Without Candidate Generation. Proc. ACM SIGMOD Int'l Conf. On Management of Data. Dallas, Texas: ACM Press, (2000) 1–12

5. Pasquier, N., Bastide, Y., Taouil, R. et al.: Discovering Frequent Closed Itemsets for Association Rules. Proc. Int'l Conf. On Database Theory. Jerusalem: Springer-Verlag, (1999) 398–416
6. Zaki, M., Hsiao, C.: CHARM: An Efficient Algorithm for Closed Itemset Mining. In Proc. SIAM Int'l Conf. on Data Mining. Arlington: SIAM, (2002) 12–28
7. Rioult, F., Boulicaut, J., Crémilleux, B. et al.: Using Transposition for Pattern Discovery from Microarray Data. DMKD'03. San Diego, CA: ACM press, (2003) 73–79
8. Pan, F., Cong, G., Tung, A. et al.: CARPENTER: Finding Closed Patterns in Long Biological Datasets. SIGKDD'03. Washington, D.C.: ACM Press, (2003) 637–642
9. Cong, G., Tan, K., Tung, A. et al.: Mining Frequent Closed Patterns in Microarray Data. ICDM'04. IEEE Press, (2004) 363–366
10. http://www.broad.mit.edu/cgi-bin/cancer/datasets.cgi
11. http://www.broad.mit.edu/cancer/pub/dlbcl

Exploring Essential Attributes for Detecting MicroRNA Precursors from Background Sequences

Yun Zheng, Wynne Hsu, Mong Li Lee, and Limsoon Wong

Department of Computer Science, School of Computing
National University of Singapore, Singapore 117543
{zhengy, whsu, leeml, wongls}@comp.nus.edu.sg

Abstract. MicroRNAs (miRNAs) have been shown to play important roles in post-transcriptional gene regulation. The hairpin structure is a key characteristic of the microRNAs precursors (pre-miRNAs). How to encode their hairpin structures is a critical step to correctly detect the pre-miRNAs from background sequences, i.e., pseudo miRNA precursors. In this paper, we have proposed to encode the hairpin structures of the pre-miRNA with a set of features, which captures both the global and local structure characteristics of the pre-miRNAs. Furthermore, we find that four essential attributes are discriminatory for classifying human pre-miRNAs and background sequences with an information theory approach. The experimental results show that the number of conserved essential attributes decreases when the phylogenetic distance between the species increases. Specifically, one A-U pair, which produces the U at the start position of most mature miRNAs, in the pre-miRNAs is found to be well conserved in different species for the purpose of biogenesis.

1   Introduction

MicroRNAs (miRNAs) are small non-coding RNAs of about 22 nucleotides long. More and more evidences show that miRNAs play important roles in gene regulation and various biological processes, as reviewed in [1,2,3]. MicroRNAs transcripts, which may be produced by RNA polymerase II or III [3], often fold to form stem loop structures, and become what are called primary miRNAs, or pri-miRNAs. In the nucleus, the Drosha RNase III endonuclease cleavages both strands of the stem at the base of the primary stem loop [4], and produce the pre-miRNAs. Then, in cytoplasm, a second RNase III endonuclease, Dicer, together with its dsRNA-binding partner protein makes a second pair of cuts and defines the other end of the mature miRNAs (see example in Figure 1), which produces the miRNA:miRNA* duplex. Finally, the miRNA stand is separated from the duplex by the helicase and form the mature miRNA molecules [2,3,4]. The mature miRNAs are then loaded to RNA-induced silencing complex (RISC), which binds the 3′ untranslate region of messenger RNAs of the miRNA target genes to repress the production of related proteins [3,5].

M.M. Dalkilic, S. Kim, and J. Yang (Eds.): VDMB 2006, LNBI 4316, pp. 131–145, 2006.
© Springer-Verlag Berlin Heidelberg 2006

The hairpin structures of the pre-miRNAs are highly conserved in different species [6,7]. Thus, how to convert the hairpin structures into informative features is a critical step to correctly identify the pre-miRNAs against the background sequences, i.e., pseudo pre-miRNAs.

There have been some endeavors for this purpose. The MirScan relied on the observation that the known miRNAs derive from phylogenetically conserved stem loop precursor RNAs with characteristic features [7]. The MiRseeker has been used to identify miRNA genes from insect DNA sequences [6]. It uses the hairpin structure predicted with the Mfold [8] as the primary criteria, but also takes into account the nucleotide divergence of miRNA candidates. The phylogenetic shadowing is a new method to find miRNA genes by comparing DNA sequences of different species [9,10]. Xue *et al.* [11] proposed a triplet-SVM classifier which encoded the hairpin structures with local structure features and obtained good sensitivity for both human data sets and data sets of other species. Bentwich *et al.* [12] proposed to score the pre-miRNAs with thermodynamical stability and structural features, which mainly capture the global properties of the hairpin structures, to classify the pre-miRNA. Sewer *et al.* [13] proposed an SVM-based method to find clustered pre-miRNAs. Yang *et al.* [14] proposed to encode the pre-miRNAs with their secondary structures, the upstream and downstream sequences.

However, first and foremost, few endeavors have been given to exploring the essential attributes for classifying pre-miRNAs and finding the biological roles of these essential attributes. Second, pre-miRNAs have been phylogenetically conserved not only for the whole hairpin structures but also for local properties at the level of nucleotides and their secondary structures, as shown in [6,15]. Finally, the specificities of the existing methods [13,11] still need to be improved.

In this research, we propose to encode the hairpin structures with the combination of global and local characteristics. In our approach, the pre-miRNA sequences and negative samples are first analyzed using the RNAfold software [16]. Second, the global characteristics of the hairpins, such as the number of base pairs and GC content, and the local structure triplet elements are used to encode the pre-miRNAs and background sequences that are predicted to contain hairpin structures with 43 features. Third, the resulted data sets are used to build classification models, which display better performance, especially specificity, for new data sets. Finally, we investigate the phylogenetically conserved essential attributes of the pre-miRNAs with the Discrete Function Learning (DFL) algorithm [17]. These features found by the DFL algorithm are accurate in predicting the pre-miRNAs, which is discussed with respect to the biogenesis mechanism of the miRNAs.

The rest of the paper is organized as follows. In Section 2, we introduce the features for encoding the pre-miRNAs and the classification algorithms for evaluating the separation ability of the generated features. In Section 3, we briefly review the DFL algorithm. In Section 4, we introduce the data sets and show the experimental results. In Section 5, we summarize the paper and discuss some future directions.

Fig. 1. The secondary structure of *Homo sapiens miR-1-1* pre-miRNA. The nucleotides in brackets represent the mature miRNA. The two triangles represent the happening of the local feature A·((at position 7 and 50. The two positions pointed by the two arrows are the cutting points of Dicer [5], which produces the miRNA:miRNA* duplex by cutting off the central loop on the right side at the two positions.

2 Methods

In this section, we show how to encode the pre-miRNA sequences and their secondary structures with a set of features which captures both their global and local characteristics. Then, we briefly review the classification algorithms used in this research.

2.1 Encoding the Hairpin Structures of Pre-miRNAs

In our approach, the secondary structures of the pre-miRNAs, as well as the candidates, are predicted with the RNAfold [16]. Then, we propose to encode the nucleotide sequences and secondary structure sequences of the pre-miRNAs with 43 features, which consist of 11 global features and 32 local features.

We first talk about the global features. The 11 global features of the pre-miRNAs are symmetric difference, number of basepairs, GC content, length basepair ratio (length of the sequence/the number of basepairs), sequence length, length of central loop, free energy per nucleotide, bulge size, bulge number, tail length and the number of tail(s).

The definitions of these features are given as follows with an example in Figure 1. To be convenient, the pre-miRNA hairpin is divided into two arms. The left arm is from 5' end (upper in Figure 1) to the center of the central loop, and the rest nucleotides form the right arm. The symmetric difference is defined as the difference of length of the two arms. For example, the symmetric difference of the *hsa-miR-1-1* precursor in Figure 1 is 2. The bulge size is defined as the size of the largest mismatch region in either of two arms. As shown in Figure 1, the largest mismatch region is the three consecutive mismatch nucleotides in the right arm. Thus, the bulge size is 3 for the *hsa-miR-1-1* precursor. The bulge number is the defined as the larger number of bulge in the two arms. Similarly, there are 3 bulges in the right arm of the *hsa-miR-1-1* precursor. Thus, its bulge number is 3. The tail length is defined as the the length of the longer free tail of the two arms. The free energy per nucleotide is obtained by dividing the free energy given by the RNAfold program with the number of nucleotide in the pre-miRNAs. For *hsa-miR-1-1* precursor, the free energy given by the RNAfold program is -30 kcal/mol. Then, the energy per nucleotide is -30/71 = -0.42 kcal. In summary, for the *hsa-miR-1-1* pre-miRNA in Figure 1, the values of the eleven global features are 2, 28, 0.37, 2.54, 71, 5, -0.42 (kcal), 3, 3, 1 and 2 respectively.

We choose these global features of the hairpins based on the following considerations. First, the pre-RNAs sequences have lower GC content than background sequences [12]. Second, the pre-miRNAs have lower folding energy than background sequences [18]. Third, we noticed that the lengths of the two arms of the pre-miRNAs are often equal or approximately equal. But the lengths of the two arms of the pseudo pre-miRNAs may be quite different. Fourth, the length of pre-miRNAs has a stable distribution [19]. Fifth, the number of basepairs and length basepair ratio are important features to decide the free energy [20]. Sixth, the number of bulge and the size of bulges are related to length basepair ratio. Seventh, we also noticed that the pre-miRNAs have no or shorter free tails at the ends of two arms than background sequences.

The local features is defined with the triplet elements proposed by Xue *et al.* [11]. One triplet is defined by one nucleotide and the secondary structure of its -1,0,+1 positions. There are 4 nucleotide, A, C, G, U, and 2 possible secondary structures, match '(' and mismatch '·'. Thus, there are totally $4 \times 2^3 = 32$ possible triplet elements. The count values of them are used as the 32 local features of our data sets. For example, for the *hsa-miR-1-1* pre-miRNA in Figure 1, the value of the feature "A·((" is 2, since it has happened at position 7 and 50, as indicated by the two dotted triangles.

2.2   The Classification Algorithms

In prior section, we demonstrate how to transform the pre-miRNA into a set of features, which carries the information of the class value of the sequences. The converted data sets are used by different algorithms to build predictors (classifiers).

In this study, we use four classification algorithms to demonstrate the value of encoding the pre-miRNA with both the global and local structural properties. The selected algorithms are the Support Vector Machines (SVM) algorithm [21], the C4.5 algorithm [22], the k-Nearest-Neighbors (kNN) algorithm [23] and the RIPPER algorithm [24].

3   The Discrete Function Learning Algorithm

To find which subset of features are relatively more important when used to predict the samples of different species, we use the Discrete Function Learning algorithm [17] to find the *essential attributes* (EAs) that contribute most to the class distinctions between samples. As to be introduced, there are two parameters for the DFL algorithm, the expected cardinality K and the ϵ value. The choice of parameters of the DFL algorithm is available in our early work [17] or at the supplementary website [1] of this paper.

---

[1] The supplements of this paper are available at `http://www.comp.nus.edu.sg/~wongls/projects/miRNA/suppl-info/vldb2006.htm`.

We will first introduce some notation. We use capital letters to represent discrete random variables, such as X and Y; lower case letters to represent an instance of the random variables, such as x and y; bold capital letters, like $\mathbf{X}$, to represent a vector; and lower case bold letters, like $\mathbf{x}$, to represent an instance of $\mathbf{X}$. In the remainder parts of this paper, we denote the attributes except the class attribute as a set of discrete random variables $\mathbf{V} = \{X_1, \ldots, X_n\}$, the class attribute as variable Y. The entropy of X is represented with $H(X)$, and the mutual information between X and Y is represented with $I(X; Y)$.

In this section, we start with a the theoretic background of information theory. Then, we introduce the motivation of the DFL algorithm. Finally, we briefly describe the DFL algorithm.

3.1   Theoretic Background

The entropy of a discrete random variable X is defined in terms of probability of observing a particular value x of X as [25]:

$$H(X) = -\sum_x P(X = x) log P(X = x).$$

The entropy is used to describe the diversity of a variable or vector. The more diverse a variable or vector is, the larger entropy it will have. Hereafter, for the purpose of simplicity, we represent $P(X = x)$ with $p(x)$, $P(Y = y)$ with $p(y)$, and so on. The mutual information between a vector $\mathbf{X}$ and Y is defined as [25]:

$$I(\mathbf{X}; Y) = H(Y) - H(Y|\mathbf{X}) = H(\mathbf{X}) - H(\mathbf{X}|Y) = H(\mathbf{X}) + H(Y) - H(\mathbf{X}, Y) \quad (1)$$

Basically, the stronger the relation between two variables, the larger mutual information they will have. Zero mutual information means the two variables are independent or have no relation.

The conditional mutual information $I(X; Y|Z)$ [26](the mutual information between X and Y given Z) is defined by

$$I(X; Y|Z) = \sum_{x,y,z} p(x, y, z) \frac{p(x, y|z)}{p(x|z)p(y|z)}.$$

The chain rule for mutual information is give by Theorem 1, for which the proof is available in [26].

Theorem 1. $I(X_1, X_2, \ldots, X_n; Y) = \sum_{i=1}^{n} I(X_i; Y|X_{i-1}, \ldots, X_1)$.

3.2   Motivation

$I(\mathbf{X}; Y)$ is evaluated with respect to $H(Y)$ in the DFL algorithm, which is different from those in existing methods, as shown in Equation 2. Suppose that $\mathbf{U}_{s-1}$ is the already selected feature subset at the step $s - 1$, and the DFL algorithm is trying to add a new feature $X_i \in \mathbf{V} \setminus \mathbf{U}_{s-1}$ to $\mathbf{U}_{s-1}$. Specifically, $X_{(1)} = argmax_i I(X_i; Y)$, and

$$X_{(s)} = argmax_i I(\mathbf{U}_{s-1}, X_i; Y), \quad (2)$$

where $\forall s$, $1 < s \leq k$, $\mathbf{U}_1 = \{X_{(1)}\}$, and $\mathbf{U}_s = \mathbf{U}_{s-1} \cup \{X_{(s)}\}$. From Theorem 1, we have

$$I(\mathbf{U}_{s-1}, X_i; Y) = I(\mathbf{U}_{s-1}; Y) + I(X_i; Y | \mathbf{U}_{s-1}). \tag{3}$$

In Equation 3, note that $I(\mathbf{U}_{s-1}; Y)$ does not change when trying different $X_i \in \mathbf{V} \setminus \mathbf{U}_{s-1}$. Hence, the maximization of $I(\mathbf{U}_{s-1}, X_i; Y)$ in the DFL algorithm is actually maximizing $I(X_i; Y | \mathbf{U}_{s-1})$, the conditional mutual information of X_i and Y given the already selected features $\mathbf{U}_{s-1}$, i.e., the information of Y not captured by $\mathbf{U}_{s-1}$ but carried by X_i.

To measure which subset of features is optimal, we restate the following theorem, which is the theoretical foundation of our algorithm. It has been proved that if $H(Y|X) = 0$, then Y is a function of X [26]. Since $I(X;Y) = H(X) - H(Y|X)$, it is immediate to obtain Theorem 2.

Theorem 2. *If the mutual information between $\mathbf{X}$ and Y is equal to the entropy of Y, i.e., $I(\mathbf{X}; Y) = H(Y)$, then Y is a function of $\mathbf{X}$.*

The entropy $H(Y)$ represents the diversity of the variable Y. The mutual information $I(\mathbf{X}; Y)$ represents the relation between vector $\mathbf{X}$ and Y. From this point of view, Theorem 2 actually says that the relation between vector $\mathbf{X}$ and Y are very strong, such that there is no more diversity for Y if $\mathbf{X}$ has been known. In other words, the value of $\mathbf{X}$ can fully determine the value of Y.

3.3   Training Methods

A classification problem is trying to learn or approximate a function, which takes the values of attributes (except the class attribute) in a new sample as input and output a categorical value which indicates the class of the sample under consideration, from a given training data set. The goal of the training process is to obtain a function which makes the output value of this function be the class value of the new sample as accurately as possible. From Theorem 2, the problem is converted to finding a subset of attributes $\mathbf{U} \subseteq \mathbf{V}$ whose mutual information with Y is equal to the entropy of Y. The $\mathbf{U}$ is the EAs that we are trying to find from the data sets. Here, we will briefly describe the main steps of the DFL algorithm as shown in the following.

1. $\forall X_i \in \mathbf{V}$, compute $I(X_i; Y)$;
2. add $A = argmax_i I(X_i; Y)$ to the EA set $\mathbf{U}_1$;
3. $\forall X_i \in \mathbf{V} \setminus \mathbf{U}_{s-1}$, compute $I(\mathbf{U}_{s-1}, X_i; Y)$;
4. add $B = argmax_i I(\mathbf{U}_{s-1}, X_i; Y)$ to the EA set $\mathbf{U}_{s-1}$;
5. repeat 3-4, until find $\mathbf{U}$ so that $I(\mathbf{U}; Y) = H(Y)$.

The DFL algorithm will find the most informative feature A in the first step. Then, the DFL algorithm will try every subsets with A and another remaining feature in $\mathbf{V}$, and find the most informative feature subset $\{A, B\}$ in the second step. Next, the similar calculation will be done until the target combination $\mathbf{U}$, which satisfies the criterion of Theorem 2, is found.

To prevent exhaustive search of all subsets of $\mathbf{V}$, one parameter called the expected cardinality K of the EAs is introduced to restrict the searching space to subsets with $\le K$ features.

After $\mathbf{U}$ is found, the DFL algorithm will stop its searching process, and obtain the classifiers by deleting the non-essential attributes and duplicate rows in the training data sets.

3.4   The ϵ Value Method

We also introduce a method called ϵ value to overcome the noisy problems [17]. Theorem 2, the exact functional relation demands the strict equality between the entropy of Y, $H(Y)$ and the mutual information of $\mathbf{X}$ and Y, $I(\mathbf{X}; Y)$. However, this equality is often ruined by the noisy data, like microarray gene expression data. In these cases, we have to relax the requirement to obtain a best estimated result. By defining a significant factor ϵ, if the difference between $I(\mathbf{X}; Y)$ and $H(Y)$ is less than or equal to $\epsilon \times H(Y)$, then the DFL algorithm will stop the searching process, and build the classifier for Y with $\mathbf{X}$ at the significant level ϵ. The ϵ is the second parameter of the DFL algorithm.

3.5   Prediction Method

After the DFL algorithm obtaining the classifiers as function tables of the pairs $(\mathbf{u}, y)$, the most reasonable way to use such function tables is to check the input values $\mathbf{u}$, then find the corresponding output values y. Therefore, we perform predictions in the space defined by the EAs $\mathbf{U}$, the *EA space*, with the 1-Nearest-Neighbor (1NN) algorithm [23] based on the Hamming distance defined as follows.

Definition 1. *Let* $1(a, b)$ *be an indicator function, which is 0 if and only if* $a = b$, *otherwise is 1. The Hamming distance between two arrays* $\mathbf{A} = [a_1, \ldots, a_n]$ *and* $\mathbf{B} = [b_1, \ldots, b_n]$ *is* $Dist(\mathbf{A}, \mathbf{B}) = \sum_{i=1}^{n} 1(a_i, b_i)$.

Note that the Hamming distance [27] is dedicated to binary arrays, however, we do not differentiate between binary or non-binary cases in this paper. We use the Hamming distance as a criterion to decide the class value of a new sample, since we believe that the rule with minimum Hamming distance to the EA values of a sample contains the maximum information of the sample. Thus, the class value of this rule is the best prediction for the sample.

In the prediction process, if a new sample has same distance to several rules, we choose the rule with the biggest count value happened in the training data set.

4   Results

In this section, we first introduce the data sets used. Then, we show the experimental results. All data sets and software used in this study are available at the supplementary website of this paper.

Table 1. The summary of data sets

	Data Set	Sample #	Class
0	TR-C (training)	163/168	pre-miRNAs/background
1	TE-C1	30	pre-miRNAs
2	TE-C2	1000	background
3	CONSERVED-HAIRPIN(T3)	2444	background
4	UPDATED(T4)	39	pre-miRNAs
5	*Mus musculusi*(mmu)	36	pre-miRNAs
6	*Rattus norvegicus*(rno)	25	pre-miRNAs
7	*Gallus gallus*(gga)	13	pre-miRNAs
8	*Danio rerio*(dre)	6	pre-miRNAs
9	*Caenorhabditis briggsae*(cbr)	73	pre-miRNAs
10	*Caenorhabditis elegans*(cel)	110	pre-miRNAs
11	*Drosophila pseudoobscura*(dps)	71	pre-miRNAs
12	*Drosophila melanogaster*(dme)	71	pre-miRNAs
13	*Oryza sativa*(osa)	96	pre-miRNAs
14	*Arabidopsis thaliana*(ath)	75	pre-miRNAs
15	*Epstein Barr Virus*(ebv)	5	pre-miRNAs
	total (1 to 15)	4094	

4.1 Data Sets and Preprocessing

In this research, we use the data sets in literature [11] to validate our approach, since it is valuable to compare the published results. These data sets are summarized in Table 1. Data set 0 to 4 is from human, and data set 5 to 15 is from other species, as indicated by their names. Data set 0 is used as the training data set, and data set 1 to 15 are used as testing data sets. There are totally 4094 samples used as testing data sets, with 3444 background sequences and 650 pre-miRNAs.

The sequences of human pre-miRNAs are obtained from miRNA registry database (release 5.0) [28]. The secondary structures of these 207 pre-miRNA sequences are predicted with the RNAfold [16]. Then, 193 sequences with only 1 loop are chosen. Next, 163 of them are randomly selected to be positive samples of the training data set, i.e., TR-C in Table 1. The rest 30 samples are used as TE-C1 testing data set.

The background sequences in data set 2 are collected from protein coding regions (CDSs) according to the UCSC refGene annotation tables [11]. The length of these sequences has the same distribution of human pre-miRNAs. The RNAfold is also used to predict the secondary structure of them. Then, the sequences with multiple loops, the sequences with less than 18 base pairs and the sequences with larger than -15kcal/mol free energy are removed. Finally, there are 8494 sequences in this data sets. Among them, 168 are randomly selected as the negative samples of the TR-C data set, and 1000, different from the 168 used, are randomly chosen as the TE-C2 testing data set.

The data set 3 also consists of background sequences, which are retrieved from the genome region from position 56,000,001 to 57,000,000 on the human chromosome 19 with the UCSC database [29]. A window of 100 nucleotides is used to scan the region and those sequences with a predicted hairpin secondary structure by the RNAfold [16] are selected. This produces 2444 background sequences in data set 3. Unlike data set 2, some sequences on data set 3 are likely to be the true pre-miRNAs. Actually, there are 3 known miRNAs (*hsa-mir-99b*, *hsa-let-7e* and *hsa-mir-125a*) in data set 3 [11].

Bentwich *et al.* [12] reported 89 new pre-miRNAs, of which 1 has multiple loops and is removed. To further remove the similar sequences, BLASTCLUST with S = 80, L = 0.5 and W = 16 is applied to the remaining 88 sequences. Only one representative sequence in each cluster is selected to remove the closely related sequences. This produces 40 non-redundant sequences, which are further checked with respect to the training data set. One of the 40 sequences that has high similarity to the training data set is removed. Finally, only 39 sequences are chosen as the data set 4.

The sequences from other species are chosen from the release 5.0 of the miRNA registry [28]. 581 out of 1138 pre-miRNAs are remained after removing the sequences with high similarity with the pre-miRNAs in the training data set. The similarity of the sequences is also calculated with BLASTCLUST with S = 80, L = 0.5 and W = 16. These pre-miRNAs form the data set 5 to 15.

4.2   Experimental Results

We have developed a software, *miREncoding*, to encode the pre-miRNA sequences, together with their secondary structure sequences, into the 43 proposed features with the Java language. We use the *Weka* software (version 3.4) [30] to evaluate the performance of the selected classification algorithms. For the SVM algorithm, polynomial kernels are used. All selected algorithms are applied to the data sets with the default settings of the *Weka* software.

To demonstrate the advantage of using both the global and local structural characteristics, we generate three data sets for each data set in Table 1 with the *miREncoding* program. The first one contains both the global and local features, the second one contains only the 32 local features, and the third one contains only the 11 global features. Then, we apply the selected algorithms to the three data sets to compare their prediction accuracies, which are shown in Figure 2.

As shown in Figure 2, the SVM, C4.5, and *k*NN algorithms show large improvements of accuracy for data set 2 and 3 when applied to data sets with all features. This suggests that the combination of global and local characteristics are critical in removing false positives, since data sets 2 and 3 are negative samples, i.e., the background sequences. For the remaining data sets, the four algorithms demonstrate stable prediction accuracies when applied to data sets with all features. When applied to only the global or local features, the prediction accuracies of the algorithms fluctuate intensively. This suggests that the combination of global or local features carries more information of the class attribute than only the global or local features.

In Figure 2 (a), we also compare the prediction performance of the SVM algorithm with the triplet-SVM classifiers in literature [11]. The SVM algorithm

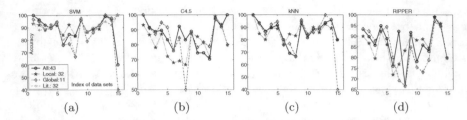

Fig. 2. The prediction accuracies of different classification algorithms, where the detailed values are available in Supplementary Table S2 to S4. (a) SVM. The curve marked with pluses represents the results of the triplet-SVM classifier, on 32 local features, in literature [11]. (b) C4.5. (c) kNN ($k = 5$). (d) RIPPER.

Table 2. The summary of prediction performance of the classification algorithms. The values shown present the performance of the classification algorithms on data sets with 32 local/11 global/all 43 features respectively. The best value for each measure (MS) is shown in bold face. The SS, SP and AC in measure (MS) column stand for sensitivity, specificity and accuracy respectively.

	MS	SVM	C4.5	kNN	RIPPER	Tri-SVM[1]
human	SS	92.8/92.8/94.2	89.9/**97.1**/94.2	94.2/95.7/94.2	91.3/92.8/94.2	92.8
(D1-4)	SP	89.5/90.3/**93.3**	80.7/88.4/89.7	81.5/86.8/88.8	81.7/89.1/87.0	88.7
	AC	89.6/90.3/**93.3**	80.8/88.6/89.8	81.7/87.0/88.9	81.8/89.2/87.1	88.8
other	SS[2]	**91.9**/89.8/91.7	84.7/85.5/83.3	87.2/90.4/87.8	89.7/84.0/87.8	90.9
species						
total	SS	**92.0**/90.2/**92.0**	85.2/86.8/84.5	88.9/90.9/89.4	89.9/84.9/88.5	91.1
	SP	89.5/90.3/**93.3**	80.7/88.4/89.7	81.5/86.8/88.8	81.7/89.1/87.0	88.7
	AC	89.9/90.3/**93.1**	81.4/88.1/88.9	85.9/87.5/88.9	83.0/88.4/87.2	89.1

[1] This column shows the results of the triplet-SVM classifier [11]. [2] The sensitivity equals to the accuracy, since there are only positive samples for data set 5 to 15.

performs better than the triplet-SVM classifiers with the more information given by the global feature of the pre-miRNAs. Especially, the total specificity is improved from 88.7% of the triplet-SVM classifiers to 93.3% of the SVM algorithm in our study, which has totally reduced 40.6% (or 158 samples) false positives in literature [11]. For the local data sets, the prediction accuracies of the SVM algorithm in our study are slightly better than those in literature [11]. We attribute this to the encoding region in our research. We encode the triplet local features for the whole pre-miRNAs, except the first and last nucleotide. However, only the paired regions of the pre-miRNAs are encoded into triplet features in [11].

The prediction performance of all selected algorithms is summarized in Table 2. As shown in Table 2, the SVM algorithm performs best for these data sets among all selected algorithms and method in literature [11]. From table 2, it is also shown that the performance of the algorithms generally becomes better when applied to the data sets with all 43 features. Especially for the specificity, the SVM, C4.5 and kNN algorithms show large improvements when applied to data sets with all 43 features. For instance, the specificity of the SVM algorithm has been dramatically improved from 89.5% for local features and 90.3% for global features to 93.3%, as shown in Table 2, which means total reduction of 131 and 103 false positive predictions respectively.

4.3   Investigating the Essential Attributes

In this section, we investigate the essential attributes for classifying pre-miRNAs and background sequences with the DFL algorithm, which has been implemented with the Java language [17]. The DFL algorithm is not designed for continuous features. Hence, we discretize the continuous features with an entropy-based discretization method [31], which has been implemented in the *Weka* software, before performing feature selection with the DFL algorithm. The discretization is carried out in such a way that the training data set is first discretized. Then the testing data set is discretized according to the cutting points of variables determined in the training data set. After that, the original continuous values of the selected features are used by other algorithms.

To find the optimal subset of EAs for the data sets, we first set the expected cardinality of the EAs K as 10. Next, we use the DFL algorithm to perform leave-one-out cross validation (LOOCV) on the training set with different ϵ values, from 0 to 0.8 with a step of 0.01. Then, we find that the DFL algorithm reaches its best prediction performance in the LOOCV when $\epsilon \in [0.12, 0.13]$ (see Supplementary Figure S2). When using all samples in training data set, the distributions of attributes are slightly different from those in LOOCV. Thus, we try the DFL classifiers obtained from a wider region of $\epsilon \in [0.1, 0.15]$. Finally, we choose the DFL classifier obtained when $\epsilon = 0.11$ because it shows overall better prediction accuracies for data set 1 to 4. In this way, a subset of 4 features, {A(((, G·((, length basepair ratio, energy per nucleotide}, is chosen as EAs for the human data sets, D1 to D4. For other data sets, the performance of the DFL algorithm is not as good as for data set 1 to 4. We attribute this to the phylogenetic distance between the species. Because fewer and fewer characteristics of the pre-miRNAs are conserved when the species become more distantly related. Thus, we try the non-empty subsets of these 4 EAs and choose those subsets on which the 1NN algorithm introduced in Section 3.5 produces the best prediction performances. The selected EAs are shown in Figure 3 (a).

We examine the phylogenetic relations between the species of the selected data sets with miRBase [32]. As shown in Figure 3 (b), data set 5 and 6 are from the Rodentia which has the closest relation with the species, *Homo sapiens*, of the training data set, D0. Then, 3 out of the 4 EAs are conserved for data set 5 and 6. For other data sets, only the 1 of the 4 EAs, A(((or length basepair ratio, is conserved. This reduction of EAs suggests that less characteristics of the pre-miRNAs are conserved when the species of data sets and *Homo sapiens* of the training data set become more distantly related.

Then, we further run the selected algorithms on the features chosen by the DFL algorithm. The prediction performance the classification algorithms is shown in Figure 4 (details available in Supplementary Table S2 and S5) and

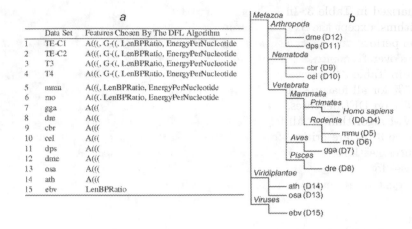

Fig. 3. The EAs chosen by the DFL algorithm. (a) The EAs chosen by the DFL algorithm. (b) The phylogenetic tree of the species of data sets from the miRBase [32].

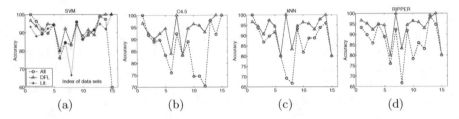

Fig. 4. The prediction accuracies of different classification algorithms. (a) SVM. The curve marked with pluses represents the results of the triplet-SVM classifier, on 32 local features, in literature [11]. (b) C4.5. (c) kNN ($k = 5$). (d) RIPPER.

Table 3. The summary of prediction performance of the classification algorithms on the DFL features. The best value for each measure is shown in bold face.

features	Measures	SVM All	DFL	C4.5 All	DFL	kNN All	DFL	RIPPER All	DFL	Tri-SVM[1]
human	sensitivity	94.2	95.7	94.2	94.2	94.2	**97.1**	94.2	95.7	92.8
(D1-D4)	specificity	**93.3**	91.5	89.7	90.8	88.8	91.7	87.0	93.1	88.7
	accuracy	**93.3**	91.6	89.8	90.9	88.9	91.8	87.1	93.2	88.8
other species	sensitivity[2]	91.7	91.9	83.3	**95.7**	87.8	95.5	87.8	95.4	90.9
total	sensitivity	92.0	92.3	84.5	**95.5**	89.4	**95.5**	88.5	95.4	91.1
	specificity	**93.3**	91.5	89.7	90.8	88.8	91.7	87.0	93.1	88.7
	accuracy	93.1	91.6	88.9	91.6	88.9	92.3	87.2	**93.5**	89.1

[1] This column shows the results of the triplet-SVM classifier [11]. [2] The sensitivity equals to the accuracy, since there are only positive samples for data set 5 to 15.

summarized in Table 3. In Figure 4 and Table 3, it is shown that the selected algorithms, except the SVM algorithm, demonstrate large improvements of prediction performance on the EAs in Figure 3 (a). For instance, the C4.5 algorithm reaches overall sensitivity of 95.5% for the EAs chosen by the DFL algorithm, as shown in Table 3. However, the C4.5 algorithm only obtains overall sensitivity of 84.5% for all features. The RIPPER algorithm reaches best overall accuracy of 93.5% on DFL features, which is slightly better than the 93.1% achieved by the SVM algorithm on all features. These results suggest that the EAs shown in Figure 3 (a) are critical for classifying the pre-miRNAs against background sequences. Although the prediction accuracies of the SVM algorithm slightly decreases for the DFL features, the SVM classifiers are much less complex than the models for all features.

5   Discussions

From Figure 3 (a) and Figure 4, it is shown that the classification algorithms are accurate on one local feature, A(((, for data sets whose species are distantly related to the species of the training data set. The A(((feature is actually originating from the A-U pairs in the pre-miRNAs. By examining the distribution of A(((in the training data set, it is known that there tend to be more A-U pairs in the pre-miRNAs than in background sequences. We attribute this higher frequency of A-U pair to two reasons. First, we consider the biogenesis process of miRNAs. It is reported that most known miRNAs begin with a U [15,19], which is originally coming from an A-U pair in the pre-miRNAs, as shown in Figure 1. In the biogenesis of the mature miRNAs, the Dicer recognizes the A-U pair in the pre-miRNAs, and performs the second cut in the biogenesis of mature miRNAs exactly at the A-U pair to produce the miRNA:miRNA* duplex [5]. This indicates that the A(((feature found in this study is critical for the biogenesis of the mature miRNAs. The high accuracies shown in Table 3 suggest that this A-U pair is well conserved in different species, even those distantly related in the phylogenetic tree, for the biogenesis of miRNAs. Second, the lower GC content in pre-miRNAs [12] partially contributes to the higher frequency of the A-U pair in the pre-miRNAs.

We have displayed how to encode the hairpin structures of pre-miRNAs with a set of features, which captures both their global and local structural properties. Different classification algorithms have shown large improvements of their prediction performance, especially the specificity, when applied to these features. This suggests that the proposed set of features have captured more information about characteristics of the hairpin structures of the pre-miRNAs than only the local features or the global features.

We have found that four EAs, with both global and local features, are critical for classifying the testing samples from the same species as the training data set. But when the phylogenetic distance between the species of the testing data sets and training data set increases, the number of EAs is reducing gradually. The selected classification algorithms generally show better prediction performance when applied to these EAs. This indicates that the pre-miRNAs of distantly related species share less common characteristics than closely related species. Therefore, to obtain better prediction performance, it is better to use the samples from the same species or closely related species as the training data set.

The false positives provide a valuable source for finding new pre-miRNAs. The improvement of specificities of the classification algorithms when applied to the combination of global and local features, as well as the EAs, can help to significantly reduce the number of putative pre-miRNA candidates, thus to save much resource for validating them.

The pre-miRNAs with multiple loops are not considered in this research. How to encode them is a valuable future direction.

Acknowledgements

This research was supported by the research grant, R-252-000-172-593, of the Institute of Infocomm Research in Singapore. We also thank Xue et al. [11] for their generous sharing of their data sets.

References

1. Alvarez-Garcia, I., Miska, E.A.: MicroRNA functions in animal development and human disease. Development **132** (2005) 4653–62
2. Ambros, V.: The functions of animal microRNAs. Nature **431** (2004) 350–5
3. Bartel, D.P.: MicroRNAs: Genomics, biogenesis, mechanism, and function. Cell **116** (2004) 281–297
4. Lee, Y., et al.: The nuclear RNase III Drosha initiates microRNA processing. Nature **425** (2003) 415–419
5. Zamore, P.D., Haley, B.: Ribo-gnome: The Big World of Small RNAs. Science **309**(5740) (2005) 1519–1524
6. Lai, E.C., et al.: Computational identification of Drosophila microRNA genes. Genome Biol **4** (2003) R42
7. Lim, L.P., et al.: Vertebrate MicroRNA Genes. Science **299**(5612) (2003) 1540–
8. Zuker, M.: Mfold web server for nucleic acid folding and hybridization prediction. Nucl. Acids Res. **31**(13) (2003) 3406–3415
9. Berezikov, E., et al.: Phylogenetic shadowing and computational identification of human microRNA genes. Cell **120** (2005) 21–24
10. Boffelli, D., et al.: Phylogenetic Shadowing of Primate Sequences to Find Functional Regions of the Human Genome. Science **299**(5611) (2003) 1391–1394
11. Xue, C., et al.: Classification of real and pseudo microRNA precursors using local structure-sequence features and support vector machine. BMC Bioinformatics **6**(1) (2005) 310
12. Bentwich, I., et al.: Identification of hundreds of conserved and nonconserved human microRNAs. Nature Genetics **37**(7) (2005) 766–70
13. Sewer, A., et al.: Identification of clustered microRNAs using an ab initio prediction method. BMC Bioinformatics **6**(1) (2005) 267
14. Yang, L., Hsu, W., Lee, M., Wong, L.: Identification of microRNA precursors via svm. In: Proc. of the 4th Asia-Pacific Bioinformatics Conference. (2006) 267–276

15. Lewis, B.P., Burge, C.B., Bartel, D.P.: Conserved seed pairing, often flanked by adenosines, indicates that thousands of human genes are microRNA targets. Cell **120** (2005) 15–20
16. Hofacker, I.L.: Vienna RNA secondary structure server. Nucl. Acids Res. **31**(13) (2003) 3429–3431
17. Zheng, Y., Kwoh, C.K.: Identifying simple discriminatory gene vectors with an information theory approach. In: Proc. of the 4th Computational Systems Bioinformatics Conference, CSB 2005, Stanford, CA (2005) 12–23
18. Bonnet, E., et al.: Evidence that microRNA precursors, unlike other non-coding RNAs, have lower folding free energies than random sequences. Bioinformatics **20**(17) (2004) 2911–2917
19. Wang, X.J., et al.: Prediction and identification of Arabidopsis thaliana microRNAs and their mRNA targets. Genome Biology **5**(9) (2004) R65
20. Zuker, M., Stiegler, P.: Optimal computer folding of large RNA sequences using thermodynamics and auxiliary information. Nucl. Acids Res. **9**(1) (1981) 133–148
21. Platt, J.: Fast training of support vector machines using sequential minimal optimization. In: Advances in kernel methods: support vector learning. MIT Press, Cambridge, MA (1999) 185–208
22. Quinlan, J.R.: C4.5: Programs for machine learning. Morgan Kaufmann, San Francisco, CA (1993)
23. Aha, D., Kibler, D., Albert, M.: Instance-based learning algorithms. Machine Learning **6** (1991) 37–66
24. Cohen, W.W.: Fast effective rule induction. In: Proc. of the 12th International Conference on Machine Learning, Tahoe City, CA, Morgan Kaufmann (1995) 115–123
25. Shannon, C., Weaver, W.: The Mathematical Theory of Communication. University of Illinois Press, Urbana, IL (1963)
26. Cover, T.M., Thomas, J.A.: Elements of Information Theory. John Wiley & Sons, Inc., New York, NY (1991)
27. Hamming, R.: Error detecting and error correcting codes. Bell System Technical Jounral **9** (1950) 147–160
28. Griffiths-Jones, S.: The microRNA Registry. Nucl. Acids Res. **32**(90001) (2004) D109–111
29. Karolchik, D., et al.: The UCSC Genome Browser Database. Nucl. Acids Res. **31**(1) (2003) 51–54
30. Frank, E., et al.: Data mining in bioinformatics using Weka. Bioinformatics **20**(15) (2004) 2479–2481
31. Fayyad, U.M., Irani, K.B.: Multi-interval discretization of continuous-valued attributes for classification learning. In: Proc. of the 13th International Joint Conference on Artificial Intelligence, IJCAI-93, Chambery, France (1993) 1022–1027
32. Griffiths-Jones, S., et al.: miRBase: microRNA sequences, targets and gene nomenclature. Nucl. Acids Res. **34**(S1) (2006) D140–144.

A Gene Structure Prediction Program Using Duration HMM

Hongseok Tae[1,2], Eun-Bae Kong[2], and Kiejung Park[1]

[1] Information Technology Institute, SmallSoft Co., Ltd., Jang-dong 59-5,
Yuseong-gu, Daejeon, 305-811, South Korea
{hstae, kjpark}@smallsoft.co.kr
http://www.smallsoft.co.kr
[2] Dept. of Computer Engineering, Chungnam National University, Gung-dong 220,
Yuseong-gu, Daejeon, 305-764, South Korea
keb@ce.cnu.ac.kr

Abstract. Gene structure prediction, which is to predict protein coding regions in a given nucleotide sequence, is a critical process in annotating genes and greatly affects gene analysis and genome annotation. As the gene structure of eukaryotes is much more complicated than that of prokaryotic genes, eukaryotic gene structure prediction should have more diverse and more complicated computational models. We have developed GeneChaser, a gene structure prediction program, using a duration hidden markov model. GeneChaser consists of two major processes, one of which is to train datasets to produce parameter values and the other of which is to predict protein coding regions based on the parameter values. The program predicts multiple genes rather than a single gene from a DNA sequence. To predict the gene structure for a huge chromosomal DNA sequence, it splits the sequence into overlapped fragments and performs prediction process for each fragment. A few computational models were implemented to detect signal patterns and their scanning efficiency was evaluated. Based on a few criteria, its prediction performance was compared with that of a few commonly used programs, GeneID and Morgan.

Keywords: HMM, Gene prediction, GeneChaser.

1 Introduction

As finding the exact positions of genes is a core process in functional genomics and comparative genomics, many kinds of gene structure prediction models have been developed to predict a set of genetic elements related with gene expression including promoter, start codon, stop codon, protein coding region and non-coding region for a genome sequence.

In early 1980s, a few approaches were developed by Shepherd[19], Fickett[7], Staden and McLachlan[23] to find protein coding regions on genome sequences based on the distribution of amino acids. As critical factors are to discriminate between coding regions and non-coding regions, many gene prediction programs have been developed using those discrimination factors such as k-tuple frequencies[4],

M.M. Dalkilic, S. Kim, and J. Yang (Eds.): VDMB 2006, LNBI 4316, pp. 146–157, 2006.

autocorrelation[13], Fourier spectra[20], purine/pyrimidine periodicity [1], and local compositional complexity/entropy[11]. GenMark[2] predicted genes on both strands simultaneously by solving a 'shadow' of coding, while the previous gene prediction programs predicted coding regions on just one strand. It makes two kinds of markov-chains, homogeneous 5th-markov chains for non-coding regions and non-homogeneous 5th-markov chains for coding regions. Glimmer[5,18], which is the most well known gene structure program for prokaryotic genome, uses k-tuples ($k \leq 8$) as sequence patterns.

While elements of a prokaryotic gene structure include promoter, start codon, stop codon, coding region and non-coding region, a eukaryotic gene structure has additional elements such as cap, poly-A, donor, acceptor, intron and exon. Since gene prediction methods for eukaryotes was started by Fields[8] and Gelfand[9] who predicted a single gene containing an initial exon and a terminal exon from a given sequence, many programs were developed, such as GeneID[10] using a hierarchical rule to calculate probability of potential exons, GeneParser[21] using neural network and dynamic programming, GenLang[6] using a linguistic method, FGENEH[22] using discriminant analysis, Morgan[17] using decision tree, Genie[12] using generalized HMM(Hidden Markov Model) and GenScan[3] using duration HMM. Among them, the algorithms using HMM have shown relatively better results. The HMM usually consists of states such as 5'UTRs, 3'UTRs, exons, introns and intergenics and uses signal patterns to find correct positions of states.

Although standard HMM is a very efficient model to predict an event structure, it is hard to represent probability models when events repeat the same state for several times. Duration HMM is an extended model of HMM to represent events staying at one state during a period[16]. As genetic elements such as exons and introns have various lengths, it is necessary to calculate the probability of a set of states with their periods.

We have developed a program called GeneChaser using signal pattern scoring models and duration HMM. It is composed of two programs, a training program to calculate parameters for gene prediction and a prediction program based on duration HMM to estimate a set of genetic elements. GeneChaser can predict a gene structure for a eukaryotic sequence and a prokaryotic sequence. Gene structure prediction for a very long DNA sequence can cause memory space to be exhausted. To predict a gene structure for a huge chromosomal DNA sequence, GeneChaser splits the sequence into overlapped fragments and performs prediction process for each fragment. And predicted genes for the overlapped regions are selected according to a selection criterion. In the performance evaluation, we used human genome sequences as training and test data to compare the result of GeneChaser with those of well known and available programs. GeneID and Morgan were used to evaluate the result of GeneChaser.

2   Method

2.1   The HMM Structure of GeneChaser

While the prokaryotic gene structure consists of promoters, start codons, stop codons, coding regions and non-coding regions, the eukaryotic gene structure has additional elements including donors, acceptors, poly-As introns and exons. Among the elements, start codons, stop codons, donors, acceptors and poly-As have definite rules such as consensus sequences. And promoters have statistical patterns on their sequences although they do not have uniform consensus. Such properties provide important clues to predict the transition points between other elements.

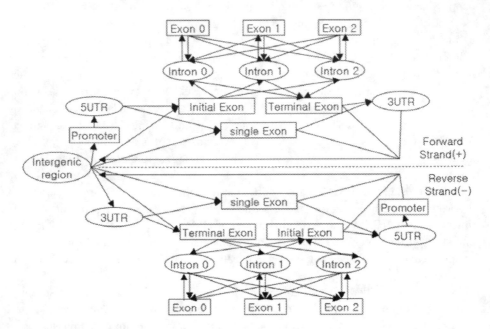

Fig. 1. The Duration HMM for the eukaryotic gene structure prediction. Rectangles and ellipses represent HMM states and arrows represent state transitions.

GeneChaser can predict both gene structures. Eukaryotes have more complicated structure than prokaryotes. The eukaryotic model of GeneChaser is depicted in Fig. 1. It consists of two programs, a training program to calculate parameters for a gene prediction process and a prediction program to produce the structure of a set of genetic elements. Parameters for gene prediction, which were designed based on duration HMM, are as follows.

    (1) Scoring matrices to search for signal positions.
      - Start codon, stop codon, promoter
      - Donor, acceptor, poly-A: **for eukaryotes**

(2) Initiation probability.
(3) Length distribution.
(4) State transition probability
(5) Segment scoring model of each state.
   - Homogeneous 5th-markov model : non-coding region
   - Non-homogeneous 5th-markov model : coding region

The duration HMM suggested in GeneChaser consists of hidden states and state transitions. For eukaryotes, 24 hidden states include promoter, 5'UTR, 3'UTR, single exon, initial exon, terminal exon, internal exon (phase 0, 1, 2), intron (phase 0, 1, 2) for both forward/reverse strand and intergenic region(Fig. 1). And the model for pro-karyotes is composed of 4 hidden states, promoter, forward coding region, reverse coding region and intergenic region. State transitions represent signals such as start codon, stop codon, donor, acceptor, promoter and poly-A, which are directional. As the promoter is too complicated to use a simple transition point, it was included in both of the state and the transition point in the model. It is one of the most distin-guished features of GeneChaser from other gene prediction program using duration HMM.

GeneChaser scans an input sequence for 6 types of signal candidates using score matrices and candidates over a cutoff value are stored. The highest probability at each candidate position is calculated using Viterbi algorithm.

The probability $r(t, i)$ that state i is observed at position t is calculated as fallows.

$$\text{for } (1 \leq t \leq T)$$
$$r(t, i) = \text{MAX}\{\pi_i * F_i (S_{1,t})*L_i(t), \text{MAX}\{r(n, Jn)*F_{Jn}(S_n,)*L_{Jn}(t\text{-}n)*M(Jn, i)\}\} \quad (1)$$
$$\text{for}(1 \leq n \leq t\text{-}1)$$
$$Jn : \text{a state at } n$$

T : the length of a input sequence.
π_i : the initial probability of state i.
$S_{1,t}$: the sequence segment from 1 to t.
n : all transition positions located before t.
Jn : the state at position n.
$F_i (S)$: the probability when segment S staying at state i.
$L_i(t)$: the length distribution when a segment staying state i with length t.
$M(Jn, i)$: the transition probability from state Jn to state i.

The process iterates until the position t reaches to T which is a total length of the sequence, and an optimal gene structure is estimated using the back-tracking algo-rithm from the position T emitting state i that shows the highest probability. If dura-tion HMM has I states and a T length, estimating optimized route using viterbi algorithm takes $O(I^2T^3)$ time complexity. The time complexity is not suitable for find-ing genes on a very long sequence. But if the length of each state can be changed within the limited range, it may prevent that calculation time increases rapidly accord-ing as a T length is prolonged.

2.2   The Training Program of GeneChaser

The training program is composed of modules to train each parameter with a training data and its detailed structure is depicted in Fig.2. GeneChaser reads CDS information as training data. After it gathers and classifies data needed to train each parameter, it reads suitable sequences to classified data. The matrix making processes of parameters are similar with each other. All processes flow two loops, the first loop extracts values from given data and sums those values, and the other loop calculates matrices using heuristic algorithm.

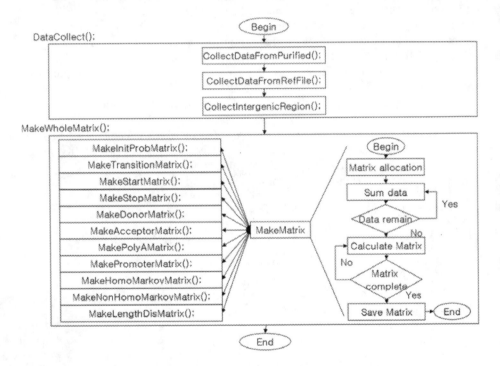

Fig. 2. The detailed structure of GeneChaser training algorithm

2.3   The Prediction Program of GeneChaser

Viterbi algorithm is used in GeneChaser to find out the optimal set of coding regions and non-coding regions constrained by signal candidates(Fig. 3). Detailed steps of the GeneChaser prediction program are as follows.

(1) Create data structures for intergenic region, promoter, 5`UTR, 3`UTR, single exon, initial exon, terminal exon, internal exon (phase 0, 1, 2), intron (phase 0, 1, 2).
(2) Create data structures for signals.
   - Start codon, stop codon, donor, acceptor, promoter, poly-A.

(3) Read state initiation probability, state transition probability, state length distribution probability, signal matrix and segment matrix from a matrix file.
(4) Read an input DNA sequence.
(5) Find all signal candidates using a signal matrix.
(6) Sort all found signals according to their positions.
(7) Using dynamic programming algorithm, evaluate all available sets of states for each signal position. The set of states with highest score is selected.
(8) The optimal set at the last base position is selected as the final set of gene prediction.

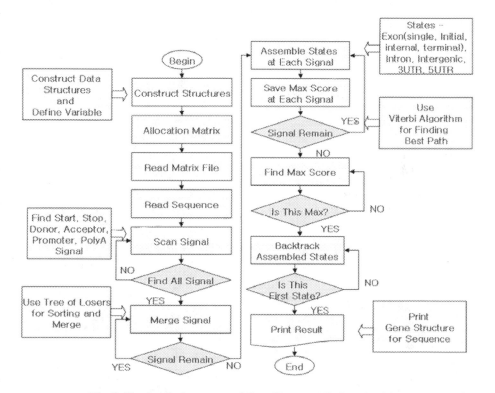

Fig. 3. The detailed structure of GeneChaser prediction algorithm

2.4 Models for Signal Scan

Signals for gene structure are divided into three kinds, one related with transcription, another related with translation, and the other related with splicing.

Promoters and poly-As are the most well-known signals related with the transcription of eukaryotic genome. GeneChaser adopts 2nd-WAM(Weight Array Matrix) for 40 base to search for promoter region including cap sites and TATA boxes. The conditional probability of nucleotide b_i at position i, $P_i = (b_i \mid b_{i-1}, b_{i-2})$ is calculated, when nucleotides b_{i-1} and b_{i-2} at positions i-1 and i-2 are given. To construct a promoter scoring matrix, -30~ 0 regions of promoter sequences from promoter databases are used. Poly-A(polyadenylation)s usually show AATAAA hexamer consensus that appears on pre-mRNA 3' end of eukaryotes. For the poly-A scanning model, WMM (Weight Matrix Model) for 6 bases of poly-A is constructed using sequences which are annotated as 'poly-A_signal' in GenBank.

Translation related signals are start and stop codons. To search for start signals, WMM is constructed with 6 bases before and 3 bases after start codons. For stop signal, 3 bases before and 6 bases after stop codons are used. Donor and acceptor are used for gene prediction as splicing signal. 3 bases before and 4 bases after GT consensuses are used to construct WMM for donor signals and 20 bases before AG consensuses are used for acceptor.

2.5 Length Distribution of States

Length distribution of states is one of the most important information to predict gene structure. For example, minimum and maximum length of a single exon is 9 and 7400 bases respectively. While length distribution of a single exon shows an Gaussian distribution between 9 bases and 400 bases, that of an intron shows geometric distribution(Fig. 4). The length distribution probability is constructed with real data without analytic formula. After splitting length data to sections with a fixed interval, the probability for each section is calculated as the number of entries in the section divided by the total entry number.

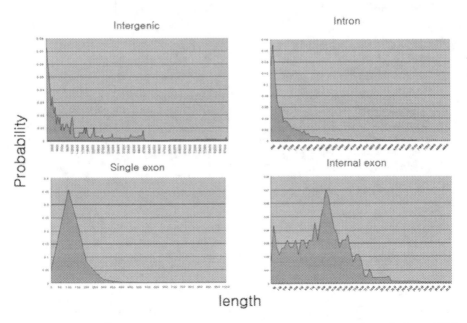

Fig. 4. Length distribution of intergenic region, intron, internal exon and single exon. While intergenic and intron show geometric distribution, single exon and internal exon show Gaussian distribution in some region.

2.6   Treatment of a Huge Sequence

While prokaryote chromosomes are a few mega bytes longs usually, most eukaryote chromosomes are longer than tens of mega bytes. Because a process of eukaryotic gene structure prediction requires memories of several tens times of the sequence length to store all information in the process, it is hard a computer to perform. To predict gene structure for a huge eukaryotic chromosomal DNA sequence, Gene-Chaser splits a huge sequence into fragments, each of which is overlapped across some hundreds bases with its adjacent fragments, and performs prediction process for each fragment(Fig. 5). To overcome the memory lack problem, GeneChaser deallo-cates the memories for predicted fragment regions after storing the predicted gene information in a gene list. If a gene overlaps with another of the gene list, longer one is selected.

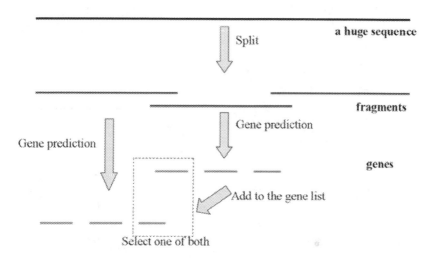

Fig. 5. The process of gene prediction for a huge sequence. GeneChaser splits a huge sequence into overlapped fragments and performs gene prediction for each fragment. When gene predic-tion is finished for one fragment, it predicted genes are added to the gene list. If a gene overlaps with another of the gene list, longer one is selected.

3   Results

3.1   Performance Comparison of GeneChaser

We used the gene prediction data of human chromosomes for performance compari-son. GeneChaser creates a set of processed data to estimate each parameter from the raw data of several sources. In order to train parameters used for GeneChaser, many kinds of training data are needed. Human chromosome 21 was used for training homogeneous markov matrix of intergenic region, initiation probability and length distribution of each state. And the dataset collected from GenBank release 89 by David Kulp at 1992 was used for homogeneous markov matrix of others and non-homogeneous markov matrix. In order to train signal scoring matrices, human promoter of EPD (Eukaryotic Promoter Database)[14] was used for promoter signal, and Kulp's dataset was used for other signals.

To test the result of GeneChaser, the Kulp's another dataset, which collected to train parameters for Genie[21] program from GenBank release 95, was used. (It is released at http://www.fruitfly.org/seq_tools/datasets/Human/). We excluded the overlapped genes with training set to avoid a bias from the test set. To test the performance of GeneChaser, GeneID and Morgan were used in comparison. Table 1 shows the accuracy of three prediction programs for each signal. All programs including GeneChaser have lower accuracy for start and stop than that for donor and acceptor signals, which means that gene prediction programs have a great difficulty in finding initial exons and terminal exons due to low consensuses around start and stop codons.

Table 1. Performance comparison of GeneChaser prediction program with other gene prediction programs with signal hit analysis.

program	signal	TP	FP	FN	Sn	Sp
GeneChaser	start	223	275	195	0.53	0.45
	stop	153	187	265	0.37	0.45
	donor	898	841	244	0.79	0.51
	accept	886	695	256	0.78	0.56
Morgan	start	172	235	245	0.41	0.42
	stop	143	264	275	0.34	0.35
	donor	761	1188	413	0.64	0.39
	accept	734	1211	377	0.66	0.37
GeneID	start	154	102	259	0.37	0.60
	stop	233	79	185	0.55	0.74
	donor	881	329	244	0.78	0.72
	accept	895	351	269	0.76	0.71

Sn = TP/Annotated signals;   Sp = TP/Predicted signals;   TP = number of true positives
FP = num of false positive; FN = number of false negative.

Table 2 shows how much the three programs predict genes correctly through the comparison of predicted exons and annotated exons. The exact match represents that two end positions for an annotated exon are the same as those of a predicted exon and partial match represents just one end of an annotated exon the same as that of predicted exon. Overlap means that no end position of an annotated exon is the same as that of predicted exon.

As GeneChaser shows 84.9% of accuracy at the total match which includes exact match, partial match and overlap, it can be used enough for practical applications.

Table 2. Performance comparison of GeneChaser prediction program with other gene prediction programs with exon hit analysis. In comparison of a predicted exon with the corresponding annotated exon, 'exact match' means that the starting and the ending sites are exactly matched, 'partial match' means that either of the starting or the ending sites is exactly matched, and 'overlap' means that neither the starting nor the ending sites is exactly matched but the compared exons are overlapped with each other.

program	# of predicted exons	exact match			# of partial match	# of overlap	total match		
		#	Sn(%)	Sp(%)			#	Sn(%)	Sp(%)
GeneChaser	2079	914	58.5	43.9	332	79	1325	84.9	63.7
Morgan	2354	631	40.4	26.8	508	130	1269	81.3	53.9
GeneID	1512	939	60.1	62.1	280	66	1285	82.4	84.9

Total number of annotated exons = 1560 Sn = (# of match / # of annotated exons) ; Sp = (# of match / # of predicted exons)

3.2 Refinement of Parameters

In the training process, all parameters are not trained well due to insufficient data. And change of some parameters affects in the result of gene prediction. We perform experiments to test how parameters affect the result by iterating gene prediction under several parameter values. According to the experimental results, signal score and segment score affected the accuracy of gene prediction more than other parameters(Fig. 6) do.

To get more precision results, gene prediction experiments were performed with several different weight values for segment and signal scores. The result showed that the change of weights for segment scores of exon and intron affected the quality of

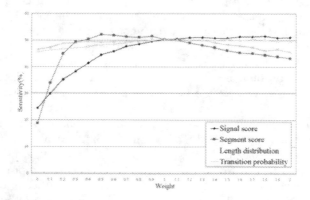

Fig. 6. Sensitivity change for the variable weights of each parameter. Sensitivity means how many real genes were found. This chart shows a segment score weight and a signal score weight affect sensitivity of gene prediction more than those of other parameter do. Weight of each score was changed from 0 to 2.0.

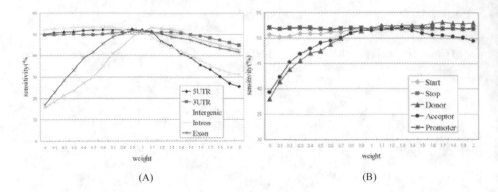

(A)   (B)

Fig. 7. Sensitivity change for the variable weights of segment and signal scores. Weights of each parameters change from 0 to 2.0. (A) Segment scores of each state affect the sensitivity of gene prediction differently. Segment scores of exon and intron affect the sensitivity more than those of others. (B) This shows that signal scores of donor and acceptor are more important factor for gene prediction than other signals.

gene prediction more than those for other parameter did(Fig. 7(A)). It means that segment scores of exon and intron were well trained, but those of other states need to be improved. In the result of analysis for signal score results, it was shown that signal scores of donor and acceptor affected the quality of gene prediction more than other signals did(Fig. 7(B)).

4 Discussion

As gene structure prediction programs are composed of several complex processes and modules to detect many genetic signals and contents, integrated knowledge and skills are required in sequence analysis to develop and improve them.

We have developed a gene structure prediction program used for both eukaryotic and prokaryotic sequences. Through the further improvement of parameter training methods and addition of more efficient computational models, GeneChaser is expected to get higher accuracy. To estimate an optimized set of states at positions of each signal candidates, we applied log odd score to GeneChaser's scoring modules for initiation probability, transition probability, length distribution, segment probability and signal probability. The method makes comparison of each score at signal positions to be easily, but spends more time to compute scores than using log score. Global scoring was applied for an analyzed sequence in dynamic programming and the accuracy of gene prediction can be affected when the sequence is very long. The combination of a global scoring method and a local scoring method for a certain sequence range is needed to overcome this weakness. Through a parameter optimization test, it is shown that a dataset of 5UTR and intergenic regions, which was used for a training process of GeneChaser, is not well adopted for gene prediction. For successful gene prediction, the training dataset of these regions should be developed.

GeneChaser program is available through the website [http://218.52.30.110:8006/~hstae/GenePrediction/index.php].

References

1. Arques, D.G. and C.J. Michel.: Periodicities in coding and noncoding regions of the genes, Vol. 143. J. Theor. Biol. (1990) 307-318.
2. Borodovsky, M. and J. McIninch.: GENMARK: parallel gene recognition for both DNA strands, Vol. 17(2). Computer & Chemistry. (1993) 123-134.
3. Burge, C. and S. Karlin.: Prediction of Complete Gene Structures in Human Genomic DNA, Vol. 268. J. Mol. Biol. 268, (1997) 78-94.
4. Claverie, J.M. and L. Bougueleret.: Heuristic informational analysis of sequences, Vol. 14. Nucl. Acids Res. (1986) 179-196.
5. Delcher, A.L., D. Harmon, S. Kasif, O. White, and S.L. Salzberg.: Improved microbial gene identification with GLIMMER, Vol. 27(23). Nucl. Acids Res. (1999) 4636-4641.
6. Dong, S. and D.B. Searls.: Vol. 23. Gene structure prediction by linguistic methods, Genomics. (1994) 540-551.
7. Fickett, J.W.: 1982. Recognition of protein coding regions in DNA sequences, Vol. 10. Nucl. Acids Res. (1982) 5503-5518.
8. Fields, C.A. and C.A. Soderlund.: gm: A practical tool for automating DNA sequence analysis, Vol. 6. Comp. Appl. Biosci. (1990) 263-270.
9. Gelfand, M.S. and M.A. Roytberg.: Prediction of the intron-exon structure by a dynamic programming approach, Vol. 30. BioSystems. (1993) 173-182.
10. Guigo, R., S. Knudsen, N. Drake, and T. Smith.: Prediction of gene structure, Vol. 226. J. Mol. Biol. (1992) 141-157.
11. Konopka, A.K. and J. Owens.: Complexity charts can be used to map functional domains in DNA, Vol. 7. Genet. Anal. Tech. Appl. (1990) 35-38.
12. Kulp, D., D. Haussler, M.G. Reese, and F.H. Eeckman.: A Generalized Hidden Markov Model for the Recognition of Human Genes in DNA, ISMB-96. (1996) 134-142.
13. Michel, C.J.: New statistical approach to discriminate between protein coding and non-coding regions in DNA sequences and its evaluation, Vol. 120. J. Theor. Biol. (1986) 223-236.
14. Périer, R.C., T. Junier, P. Bucher.: The Eukaryotic Promoter Database EPD. Vol. 26. Nucl. Acids Res. (1998) 353-357.
15. Pruitt, K.D. and D.R. Maglott.: RefSeq and LocusLink: NCBI gene-centered resources. Vol. 29(1). Nucl. Acids Res. (2001) 137-140.
16. Rabiner, L.R.: A tutorial on Hidden Markov Models and selected applications in speech recognition, Vol. 77(2). Proc. IEEE. (1989) 257-285.
17. Salzberg, S., A.L. Delcher, K.H. Fasman, and J. Henderson.: A Decision Tree System for Finding Genes in DNA, Vol. 5(4). J. Comp. Biol. (1998) 667-680.
18. Salzberg, S.L., A.L. Delcher, S. Kasif, and O. White.: Microbial gene identification using interpolated Markov models, Vol. 26(2). Nucl. Acids Res. (1998) 544-548.
19. Shepherd, J.C.W.: Method to determine the reading frame of a protein from the purine/pyrimidine genome sequence and ist possible evolutionary justification, Vol. 78. Proc. Natl. Acad. Sci. USA. (1981) 1596-1600.
20. Silverman, B.D. and R. Linsker.: A measure of DNA periodicity, Vol. 118. J. Theor. Biol. (1986) 295-300.
21. Snyder, E.E. and G.D. Stormo.: Identification of protein coding regions in genomic DNA, Vol. 248. J. Mol. Biol. (1995) 1-18.
22. Solovyev, V.V., A.A. Salamov, and C.B. Lawrence.: 1994. Predicting internal exons by oligonucleotide composition and discriminant analysis of spliceable open reading frames, Vol. 22. Nucl. Acids Res. (1994) 5156-5163.
23. Staden, R. and A.D. McLachlan.: Codon preference and its use in identifying protein coding regions in long DNA sequences, Vol. 10. Nucl. Acids Res. (1982) 141-156.

An Approximate de Bruijn Graph Approach to Multiple Local Alignment and Motif Discovery in Protein Sequences

Rupali Patwardhan[1], Haixu Tang[1,2], Sun Kim[1,2], and Mehmet Dalkilic[1,2]

[1] Center for Genomics and Bioinformatics, Indiana University,
1001 E. 3rd Street, Bloomington, IN 47405
[2] School of Informatics, Indiana University,
901 E. 10th Street, Bloomington, IN 47408
{rpatward, hatang, sunkim2, dalkilic}@indiana.edu

Abstract. Motif discovery is an important problem in protein sequence analysis. Computationally, it can be viewed as an application of the more general multiple local alignment problem, which often encounters the difficulty of computer time when aligning many sequences. We introduce a new algorithm for multiple local alignment for protein sequences, based on the de Bruijn graph approach first proposed by Zhang and Waterman for aligning DNA sequence. We generalize their approach to aligning protein sequences by building an approximate de Bruijn graph to allow gluing similar but not identical amino acids. We implement this algorithm and test it on motif discovery of 100 sets of protein sequences. The results show that our method achieved comparable results as other popular motif discovery programs, while offering advantages in terms of speed.

Keywords: Motif discovery, local alignment, de Bruijn graph, proteins.

1   Introduction

As an important problem in bioinformatics, motif discovery in protein sequences can be viewed as a direct application of a more general multiple local alignment problem. Programs like MEME [1], Gibbs Sampler [2] and BlockMaker[3] represent popular solutions to this problem and have been commonly used in many applications. However, these programs share the same shortcoming of the computational efficiency, preventing their applications to large scale cases.

Zhang and Waterman recently introduced a fundamentally novel algorithm for local multiple alignment of DNA sequences [4]. This was an extension of their EulerAlign approach for global multiple alignment [5]. They aim to construct a consensus pattern that is most consistent with all input sequences from a de Bruijn graph built from the input sequences. The consensus is then used as a query to locate all instances of the pattern. Their algorithm has the obvious advantage over the previous methods that it reduces the time complexity of the problem to approximately linear, thus significantly reducing the computational time when the input size is large.

M.M. Dalkilic, S. Kim, and J. Yang (Eds.): VDMB 2006, LNBI 4316, pp. 158–169, 2006.

In spite of its success with DNA sequences, it is not straightforward to apply this method to local multiple alignment of protein sequences. There are several important distinctions between homologue protein and DNA sequences. First, protein sequences use a larger alphabet (with 20 amino acids) than DNA sequences (with 4 nucleotides). Second, homologous protein sequences have lower sequence identity than DNA sequences. For instance, even closely homologous protein sequences may have only 50% ~ 60% identity; let alone the fact that some distantly homologous protein sequences may have as low as 30% identity. As a result, the possiblity of finding identical k-mers that would form the basis of the consensus path is much lower. Finally, amino acid residues, represented by 20 letters in protein sequences, are not equally similar with each other. [1] For example, leucine residue is more similar to other hydrophobic amino acids, e.g. isoleucine or valine, than the hydrophilic residues. These differences are represented in amino acid similarity matrices, such as PAM [6] or BLOSUM [7].

The above three differences complicate the application of de Bruijn graph approach to the protein sequence alignment. In this paper, we attempt to address these issues by introducing the concept of *approximate de Bruijn graph*. Then we adopt a similar approach as Zhang and Waterman to perform the local multiple sequence alignment by traversing this graph. We first find a heaviest path in the approximate de Bruijn graph using a heuristic greedy algorithm and deduce a consensus protein sequence from this path. Next we align this consensus sequence to each of the input sequence and finally construct the multiple alignment from these pairwise alignments. We then repeat this to find other patterns.

We applied this method to motif discovery in protein sequences. The algorithms were tested on protein families corresponding to PROSITE [8] patterns, and also compared against two other existing motif discovery algorithms, MEME [1] and PRATT [9]. Results show our algorithm outperforms PRATT and generates comparable results to MEME, but runs much faster. We also present a modified version of this algorithm, which attempts to account for motifs that contain certain non-specific positions or gaps. In case of motifs with no continuous conserved amino acid residues but with at least one occurrence of alternate conserved residues, our algorithm outperformed both MEME and PRATT.

2   Approximate de Bruijn Graph

Consider n protein sequences as input. A k-tuple is a subsequence of length k in one of the sequences. In de Bruijn graph [10], we represent each k-tuple by two connected nodes, in which each node represents a $k-1$-tuple and these two $k-1$-tuples overlap with $k-2$ letters in the original sequence; the connecting edge represents the k-tuple pointing from the prefix $k-1$ tuple to the suffix $k-1$

---

[1] Letters in DNA sequences, i.e. nucleotides, are not equally similar with each other either, as the mutation rate of transition, i.e. the mutation from purine to purine, or from pyrimidine to pyrimidine, and transversion, i.e. the mutation from purine to pyrimidine or *vice versa* differs. But the difference is subtle and ignored in DNA sequence alignment.

tuple (Fig. 1 (b)). Two identical k-tuples in the input sequences correspond to the same edge in the de Bruijn graph. The *weight* of each edge is defined as the number of identical k-tuples in the input sequences. The de Bruijn graph of the input sequences can be constructed in a progressive way. First we process one input sequence at a time and add the k-tuples in this sequence into the graph. If the incoming k-tuple corresponds to an edge existing in the current graph, we increase the weight of this edge by 1; otherwise we create a new edge (and nodes if necessary) accordingly and assign weight 1 to this edge. It is easy to show that this construction procedure is independent of the order in which input sequences are processed.

In order to take into account the similarity between amino acid residues, we define the *approximate de Bruijn graph*, which generalizes the classical definition of de Bruijn graph. Given n protein sequences, *approximate* de Bruijn graph has the same topology (nodes and edges), but different edge weights as compared to the *classical* de Bruijn graph. If one k-tuple is *similar* to w other k-tuples in the de Bruijn graph, the weight of the corresponding edge in approximate de Bruijn graph is incremented by the sum of weights of all these w k-tuples scaled by the degree of similarity and a similarity constant. If we adopt the rigorous criteria in defining the *similarity* between two k-tuples, *approximate* de Bruijn graph is reduced to the *classical* de Bruijn graph. But generally we want to relax this rigorous condition to account for amino acid similarities. We define two k-tuples a and b are *similar*, if the substitution score for each pair of corresponding residues in the two subsequences represented by two edges is larger than a predefined threshold θ. The substitution score between each pair of residues can be extracted from any standard protein similarity matrices such as BLOSUM or PAM or any other user defined similarity matrix.

3   Algorithm for Motif Discovery

Given n protein sequences, our algorithm for motif finding involves four major steps:

1. Build the classical *de Bruijn graph* from the set of input sequences;
2. Transform the *classical* de Bruijn graph into *approximate* debruijn graph by adjusting the weights of edges in the graph taking into account the amino acid similarities;
3. Traverse the *approximate de Bruijn graph* to discover significant consensus subsequences;
4. Locate all instances of the reported subsequences identified in the previous step and output their multiple alignment;

In the next section we will elaborate on these steps.

3.1   Building the Classical de Bruijn Graph

Let each edge in the graph correspond to a subsequence of length k. We adopt the following progressive procedure to build the *classical* de Bruijn graph from a set of given protein sequences.

We process the sequences one at a time, and for each sequence,

1. Move along the sequence shifting one amino acid position at a time;
2. At each position consider a subsequence of length k starting at that position;
3. Check if the edge corresponding to this subsequence already exists in the graph;
4. If it does, then increment the weight of that edge by 1; otherwise, create a new edge and assign it a weight of 1.

3.2 Adjusting Edge Weights

We define two edges to be *similar* if every pair of corresponding residues in the two subsequences represented by two edges has a positive substitution score as per a similarity matrix. This similarity matrix can be a standard substitution scoring matrix such as BLOSUM62 (used as default in this study) or any other user defined similarity matrix.

Consider two edges X and Y. They will be considered to be *similar* if and only if

$$s(X_i, Y_i) > \theta, \ \forall_{i:1 \leq i \leq k} \tag{1}$$

Where,
$X_i = i^{th}$ amino acid residue in the k-tuple represented by the edge X,
$Y_i = i^{th}$ amino acid residue in the k-tuple subsequence represented by the edge Y,
$s(X_i, Y_i) = $ substitution score for residues X_i and Y_i as per the substitution matrix.
$\theta = $ residue similarity threshold 0 as default

The weights of similar edges are incremented proportional to their similarity. For this, we introduce a relative similarity score as a measure of similarity between edges. A *relative similarity score r* between two edges X and Y is defined as,

$$r(X, Y) = \sum_{i=1}^{k} \frac{s(X_i, Y_i)}{s(X_i, X_i)} \tag{2}$$

To incorporate similarity into the graph, for each edge in the graph,

1. Check if any other edges are similar to it.
2. If yes, update the weight of this edge by adding to it the weights of these similar edges scaled by their similarity score and a user defined similarity constant (default 0.5);otherwise keep the weight unchanged.

For example, suppose edge A is under consideration. We find that edges B, C and D are similar to it. Then we update the weight of edge A as follows,

$$W'_A = W_A + K \times (r(A, B) \times W_B + r(A, C) \times W_C + r(A, D) \times W_D) \quad (3)$$

Where,
W_X = weight of edge X in classical de Bruijn graph,
W'_X = weight of edge X in approximate de Bruijn graph,
K = similarity constant

At the end of this step, we construct the approximate de Bruijn graph. The topology of this graph is the same as that of the classical de Bruijn graph. Only the weights of some edges are altered. We note that as in case of a classical de Bruijn graph, the weights as per this scheme are independent of the order in which the edges are processed. The similarity constant K can be used to adjust the effect of similar edges. Thus, if $K = 0$, the weights of edges in approximate ·de Bruijn graph is the same as in classical de Bruijn graph.

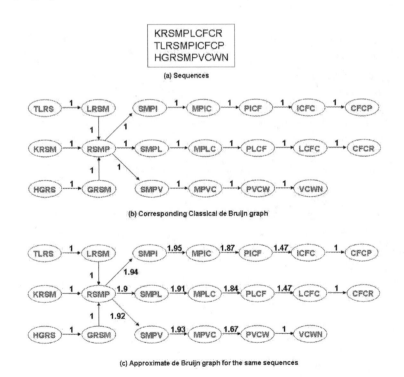

Fig. 1. *Classical* (b) and *approximate* (c) de Bruijn Graphs of a given set of protein sequences (a)

A small example shown in Fig.1 illustrates the difference in edge weights before and after similarity adjustment, and how the new weights help in identification of the motif. Figure 1(a) shows three input protein sequences. Figure 1(b) represents the classical de Bruijn graph built from these sequences. Figure 1(c) illustrates the graph after weights have been adjusted to make it *approximate*.

3.3   Traversing the Graph

A high weight for an edge in de Bruijn graph indicates that the subsequence represented by the edge and its similar subsequences occur many times in the input sequences. Therefore, we attempt to identify the highly weighted (*heavy*) paths by traversing the graph:

1. Identify the heaviest edge. The k-tuple represented by this edge is used as a seed of the consensus.
2. Starting from this seed, successive edges in the graph are traversed. At each step, the seed is extended to include the edge being traversed. Motif extension continues until the weight of the traversed edge is below a threshold value i.e. a percentage (defined by the user, default 80%) of the maximum weight of all edges in the initial approximate de Bruijn graph. The same procedure is applied to both sides of the seed. Each time there is more than one edge to traverse, we choose the one with higher weight. We then extract a consensus motif from the final path.
3. Using this consensus, identify all the occurrences of the consensus motif in the input sequences;
4. Decrease the weights of edges in the identified path [2];
5. The process is repeated until all seed-containing paths with a weight above the threshold are identified. Again this threshold is a percentage of the global maximum weight edge, but may be different from the threshold for seed extension used above.

3.4   Building Multiple Local Alignment

Once all the consensus sequences are identified, we use a linear space implementation of the Smith-Waterman local pairwise alignment [11],[12] available as a part of the FASTA package to align the consensus with each input sequence. The resulting pairwise alignments are then coverted to a local multiple alignment.

4   Algorithm for Gapped Motif Discovery

The above algorithm is very effective when the anticipated motif contains several continuous conserved residues that correspond to a highly weighted path in the approximate de Bruijn graph. However, the method may fail if the motif consists of alternating conserved and non-conserved residues, e.g. AxAxCxDx-AxGxC ('x' represents a non-specific position and may be any kind of residue)

---

[2] This is similar to the declumping procedure described by Zhang and Waterman [4]. In our case, weights of similar edges should also be appropriately adjusted

or AxCDxGxRGxC. As there is no restriction on 'x's, residues in this position in different motif instances could have negative substitution scores with each other, thus may not be over the residue similarity threshold θ. In such cases, each occurrence of the motif leads to separate non-similar edges in the graph, and hence their total weight is not high enough to be detected using a reasonable similarity threshold. Lowering θ obviously is not a good option because it will then increase false positive identifications. To address this issue, we came up with a slightly modified *gapped* version of the algorithm. The only difference between this new algorithm and the algorithm described above is in the first step of building the graph. In the new algorithm, we create two types of nodes, in addition to the nodes for the k-tuples in the input sequences, we also create nodes for *masked* (by *non-specific masks*) tuples.

4.1 Masking Subsequences

Let '1' represent a conserved amino acid and '0' represent a gap or *non-specific position*. A *mask* can then be defined as a string of 1s and 0s of length $k - 1$. All permutations and combinations of 1's and 0's will yield 2^{k-1} such masks. However, masks with more zeros than ones will result in nodes that are too non-specific. So we only consider masks with number of 1's equal to or greater than the number of 0's. The masking operation can be considered analogous to a logical bit-wise AND. The residue that overlaps with 1 is retained while the residue corresponding to 0 is replaced by a 'x'. For example, if we apply the masking operation (denoted as '*') to the subsequence, ANCD, then ANCD * 1001 = AxxD, ANCD * 1101 = ANxD, ANCD * 1011 = AxCD and so on. Each subsequence in the input sequences will be masked using all pre-designed masks, and each resulting masked subsequence will be represented by a separate node in the graph (Figure 3).

4.2 Linking Edges

Edges are linked between nodes with overlapping subsequences as before. However the overlapping is defined more flexibly because of the 'x's, leading to a much more connected graph. As an example, consider 2 nodes of subsequences ANCD and RCDE. These nodes will not be linked in the approximate de Bruijn graph. But after masking, they will be linked by an edge. This is because, after masking ANCD with 1011, we get AxCD and this now can be connected to xCDE which is obtained by masking RCDE with 0111. Suppose we have three input sequences with subsequences as shown in Figure 2(a), they clearly contain a motif AxCDx. Accordingly, instead of three different edges each with a low weight (Figure 2(b)), one collapsed edge with a higher weight will be left (Figure 2(c)) after masking.

The masking procedure enables us to detect motifs that could not be detected using the regular algorithm described previously. This is illustrated in Fig.3. Assume AxCDxxGH is the motif. Each instance of the motif will contain

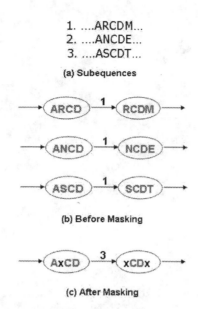

1. ....ARCDM...
2. ....ANCDE...
3. ....ASCDT...

(a) Subequences

(b) Before Masking

(c) After Masking

Fig. 2. Edge consolidation by masking subsequences in approximate de Bruijn graph. Three edges linked before masking (b) between input subsequences (a) collapse into one single edge with higher weight after masking (c).

a non-specific residue at the 'x' position. Before masking, they would lead to separate edges each with low weight in the graph. Masking allows these instances to collapse into same edges, resulting higher weights to be detected by graph traversing (highlighted in red).

5   Results

We implemented both algorithms described above in Perl and tested the program on the motif discovery from sequences of the same protein families. Our program can be accessed through a website
http://biokdd.informatics.indiana.edu/rpatward/debruijn/project.html.

5.1   Test Datasets

We used a benchmarking data set for which the optimal motif is known and manually verified. PROSITE is a well known database of motif patterns, and their related protein families [8]. The PROSITE patterns are hand curated and correspond to sites that are proven to be biologically significant, and have already been used in the past by researchers to test the performance of motif discovery algorithms. In their paper, Hart et al [13] give a justification of the relevance of using overlap with PROSITE pattern sites as a good measure for estimating the quality and accuracy of a detected motif. Thus, we used protein

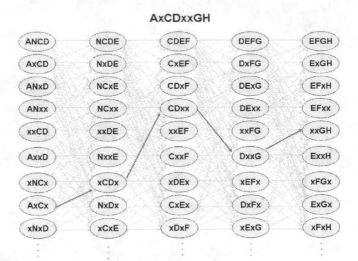

Fig. 3. An schematical example illustrating the advantage of masking procedure. After masking, motif instances (of AxCDxxGH) collapse into same edges, resulting a high weighted path that can be identified by graph traversing (highlighted in red).

families corresponding to PROSITE patterns as input, and the agreement with known PROSITE patterns as a measure of success or failure of the algorithm. We compared the performance of our algorithms with two other motif discovery algorithms, MEME and PRATT. We select these two programs since they represented two classes of conventional motif discovery methods: probabilistic (MEME) and combinatorial methods (PRATT). We note that it is likely that a motif output by either of the algorithms could be a true motif even if it is not the same one as the PROSITE pattern. But since it is not easy to ascertain this, we decided to consider just those that correspond to the respective PROSITE patterns as true motifs.

We used two different datasets. The first one consisted of a hundred protein families corresponding to first 100 PROSITE patterns in the database The second dataset (referred to as "hard-to-find" motif sets) included families corresponding to PROSITE patterns that had no continuous conserved amino acid residues, but at least one occurrence of alternating conserved residues.

As mentioned above, we first tested all programs on the first data set, i.e. protein families corresponding to first 100 PROSITE patterns that had a Data Bank Reference (PS00010 to PS00119). The families that were not included were those corresponding to 10 PROSITE entries PS00013, PS00015, PS00016, PS00017, PS00029, PS00038, PS00040, PS00043, PS00044 and PS00107. Entries PS00013 and PS00015 are rules, not patterns. PS00016, PS00017 and PS00029 are categorized as too non-specific. PS00038, PS00040, PS00043 and PS00044 do not have a PROSITE entry associated with them at present. PS00107 was excluded because it is very non-specific; there more than one thousand sequences matching this pattern.

All four programs were run with default parameters. For MEME and PRATT, since we need to specify the number of motifs, the top three motifs were considered. It can be seen that our algorithms achieve comparable performance as the other popular algorithms. As we will show below, our program, however, runs much faster.

Table 1. Results for Test Dataset I

Algorithm	No. of families
Approximate de Bruijn	69
Gapped de Bruijn	75
MEME	80
PRATT	61

On the second data set, our gapped de Bruijn algorithm performed better than both MEME and PRATT, with the results agreeing with the PROSITE pattern for 128 (79%) of the 162 families (Table 2).

Table 2. Results for Test Dataset II

Rank of motif	Gapped de Bruijn	MEME	PRATT
1	68	50	77
2	26	26	13
3	12	10	4
4	12	6	6
5	10	10	5
Total	128	102	105

We note that both of our methods have their own advantages. The conventional method can detect similar segments even if they have no exact residues in common; however all the residues in the core motif should have similarity scores greater than the threshold θ. On the other hand, the second (masked) method requires at least a couple of exactly conserved residues though they need not be continuous.

As an example consider the protein family correspoding to PS00050. The PROSITE signature for this family is $[RK](2) - [AM] - [IVFYT] - [IV] - [RKT] - L - [STANEQK] - x(7) - [LIVMFT]$. The conventional approximate de Bruijn method can detect this motif, while the gapped method version cannot. But for another PROSITE pattern, PS00118 (signature $C - C - \{P\} - x - H - \{LGY\} - x - C$), the approximate method fails to identify this motif because the motif does not have enough conserved residues to form high weight edges, whereas the gapped method can successfully identify it.

To compare the speed of four programs, we plotted the running times of all four algorithms for a range of input sizes (Figure 4). Note that we consider the time of MEME running in parallel mode using 8 processors on IBM SP cluster whereas those of the other programs running on a single processor computer, for it runs significantly slower than the other programs.

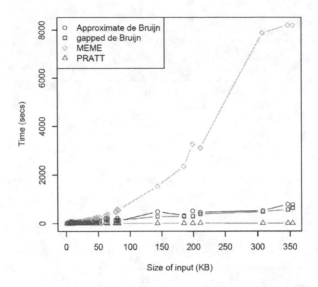

Fig. 4. Comparison of running times of four methods on data sets with different input size (X-axis)

6   Discussion

The results of our new methods look overall encouraging. It is competitive in terms of both speed as well as accuracy in comparison to existing algorithms. Furthermore, it scales very well for large inputs.

An elegant feature of the de Bruijn graph approach is that it is incremental. So in theory, even if you add one more input sequence to a precomputed data set, there is no need to start building the graph from scratch. It can be easily added to the existing graph. Also, the graph structure and weights of edges and hence the result is independent of order of input sequences.

We used the BLOSUM or PAM substitution matrices to define the similarity of amino acid residues. In principle, any kind of user defined metric for similarity could be used for this purpose. This gives the users the ability to include their biological or intuitive knowledge in the motif discovery process.

When applied to motif discovery, our algorithm eliminates the need to specify the length and/or number of motifs in advance. All motifs that are above the specified motif threshold will be reported. However, unlike a lot of enumeration algorithms, this number is usually low so that the user does not have to sift through a long list of false positives to validate the real motifs.

In addition to motif discovery, a modification of our algorithm could be applied to protein classification, i.e. to determine the familyship of a protein sequence. This could be done by first building a de Bruijn graph from the sequences known to be in each protein family and then trying to traverse the graph using the query sequence. If the sequence belongs to this family, many of the nodes and edges required for traversing this sequence will already be present in the graph. On the other hand, if hardly any of the required edges exist in the graph, the sequence is apparently quite different from the sequences in the family and hence it is unlikely to belong to the family.

Acknowledgements. We thank members of the Bioinformatics Research Group at the School of Informatics, and the members of the Bioinformatics Team at Center for Genomics and Bioinformatics for many helpful discussions. Kim was partially supported by NSF CAREER DBI-0237901.

References

1. Bailey, T.L., Elkan, C.: Fitting a mixture model by expectation maximization to discover motifs in biopolymers. In: Proceedings of the Second International Conference on Intelligent Systems for Molecular Biology, AAAI Press, Menlo Park, California (1994) 28–36
2. Lawrence, C., Altschul, S., Bogouski, M., Liu, J., Neuwald, A., Wooten, J.: Detecting subtle sequence signals: A gibbs sampling strategy for multiple alignment. Science **262** (1993) 208–214
3. Henikoff, S., Henikoff, J.G., Alford, W.J., Pietrokovski, S.: Automated construction and graphical presentation of protein blocks from unaligned sequences. Gene **163** (1995) GC17–GC26
4. Zhang, Y., Waterman, M.S.: An Eulerian path approach to local multiple alignment for DNA sequences. PNAS **102** (2005) 1285–1290
5. Zhang, Y., Waterman, M.S.: An eulerian path approach to global multiple alignment for dna sequences. Journal of Computational Biology **10** (2003) 803–819
6. Dayhoff, M., Schwartz, R., Orcutt, B.: A model of evolutionary change in proteins. In: Atlas of Protein Sequence and Structure. Volume 5(3). National Biomedical Research Foundation (1978) 345–352
7. Henikoff, S., Henikoff, J.: Amino Acid Substitution Matrices from Protein Blocks. PNAS **89** (1992) 10915–10919
8. Falquet, L., Pagni, M., Bucher, P., Hulo, N., Sigrist, C., Hofmann, K., Bairoch, A.: The prosite database, its status in 2002. Nucleic Acids Res. **30** (2002) 235–238
9. Jonassen, I.: Efficient discovery of conserved patterns using a pattern graph. CABIOS **13** (1997) 509–522
10. van Lint, J., Wilson, R.: A Course in Combinatorics. 2nd edn. Cambridge University Press (2001)
11. Myers, E.W., Miller, W.: Optimal alignments in linear space. CABIOS **4** (1988) 11–17
12. Smith, T., Waterman, M.: Identification of common molecular subsequences. Journal of Molecular Biology **147** (1981) 195–197
13. Hart, R., Royyuru, A., Stolovitzky, G., Califano, A.: Systematic and fully automated identification of protein sequence patterns. Journal of Computational Biology **7(3-4)** (2000) 585–600

Discovering Consensus Patterns
in Biological Databases

Mohamed Y. ElTabakh[1], Walid G. Aref[1],
Mourad Ouzzani[2], and Mohamed H. Ali[1]

[1] Dept. of Computer Science, Purdue University, West Lafayette IN 47906, USA
[2] Cyber Center, Purdue University, West Lafayette IN 47906, USA
{meltabak,aref,mourad,mhali}@cs.purdue.edu

Abstract. Consensus patterns, like motifs and tandem repeats, are highly conserved patterns with very few substitutions where no gaps are allowed. In this paper, we present a progressive hierarchical clustering technique for discovering consensus patterns in biological databases over a certain length range. This technique can discover consensus patterns with various requirements by applying a post-processing phase. The progressive nature of the hierarchical clustering algorithm makes it scalable and efficient. Experiments to discover motifs and tandem repeats on real biological databases show significant performance gain over non-progressive clustering techniques.

1 Introduction

A consensus pattern is a highly conserved pattern with very few substitutions where no gaps are allowed within the pattern. Discovering consensus patterns has several applications especially in biological databases for the case of motifs and tandem repeats. Motifs are highly conserved patterns that appear in the upstream region of genes. Motifs are regulatory elements that regulate the expression of genes and hence the functionality of the cell. Tandem repeats are highly conserved patterns too. They appear several times after each other in a DNA sequence. The regions in which tandem repeats appear are called *repeat regions*. Tandem repeats are considered DNA signatures, and have an important evolutionary role [30]. Discovering motifs and tandem repeats in biological databases is crucial for understanding the genetics of the cell. Discovering consensus patterns raises several challenges that make applying data mining tools such as clustering techniques a nontrivial task. In particular, the patterns that we are looking for are usually unknown, the length of the consensus patterns is unknown.

In this paper, we propose a progressive hierarchical clustering technique, called Bio-CP, for discovering consensus patterns. Bio-CP discovers consensus patterns by clustering similar patterns together over a range of fixed lengths. Each cluster represents a candidate set of consensus patterns. The processing of Bio-CP is divided into phases. In each phase, we discover the candidate consensus patterns for a certain length. Then, we proceed to the next phase in an incremental manner to obtain the candidate patterns with the subsequent length. At the end of each

M.M. Dalkilic, S. Kim, and J. Yang (Eds.): VDMB 2006, LNBI 4316, pp. 170–184, 2006.

phase, we perform a post-processing phase to apply any domain specific requirements over the candidate patterns. Bio-CP is applicable to a wide range of applications since it allows any domain specific requirements to be applied independently in a post-processing phase. Furthermore, Bio-CP executes progressively and hence significantly reduces the processing overhead compared to non-progressive clustering techniques. However, in its original form, Bio-CP involves a high overhead in the first phase due to computing and storing a large distance matrix. To address this issue, we propose several scalability techniques to reduce both the storage and CPU overheads.

The rest of the paper is organized as follows. We discuss the related work in Section 2. In Section 3, we present Bio-CP concepts and methods. Scalability issues are discussed in Sections 4. The experimental results are presented in Section 5. We conclude in Section 6.

2   Related Work

Pattern similarity is studied in several application domains. In data mining, frequent pattern mining is the problem of discovering similar patterns that appear a number of times above a certain threshold in the database, e.g., [2,4]. Frequent pattern mining techniques cannot handle efficiently the problem of discovering consensus patterns for the following reasons: (1) Consensus patterns allow approximate matching, whereas frequent pattern mining techniques (even the techniques that allow gaps) usually search for only exact matches. (2) Frequent pattern mining techniques involve high overhead in the early phases in which too many short frequent patterns are discovered. The length of these short patterns can be out of the interesting range of the consensus patterns. Similarity search techniques aim at searching for a query string in a database of sequences, e.g., [1,3,5,17]. Several data structures are developed for searching string and sequence data, e.g., suffix trees, e.g., [28,29], and suffix arrays, e.g., [27,29]. While these techniques and data structures are related to our targeted problem, they cannot be applied directly for discovering consensus patterns since we do not have a query string to search for in the first place.

Clustering techniques rely on grouping similar patterns or objects together [10,14]. MOPAC [13] is an agglomerative clustering technique to discover motif consensus patterns in biological databases. MOPAC solves the problem for a specific motif length. For a length range, MOPAC needs to be re-executed for each candidate motif length. Usually, using existing clustering techniques to discover consensus patterns is limited to discovering such consensus for a specific length, which is not general enough since the length of such patterns is not known a priori. In contrast, our techniques extend clustering techniques to work over a range of lengths in a scalable way.

Several statistical and non-statistical approaches have been proposed in biological databases. Most of these approaches target either motifs or tandem repeats but not both. Examples of statistical techniques for discovering motif patterns can be found in [7,8,16,20]. A statistical technique for finding tandem repeats in DNA sequences is proposed in [9]. Other non-statistical techniques includes PROJECTION [11], COPIA [21], and WINNOWER [24] for discovering motifs and one technique [19,22] for discovering tandem repeats.

3   Bio-CP Concepts and Methods

Bio-CP is a progressive technique for discovering consensus patterns for a given database of sequence, a length range, and a threshold for clustering patterns together at a certain length. More precisely, given (1) a database D of N sequences, i.e., $D=\{S_1, S_2,..., S_N\}$ where each sequence S_i ($1\leq i\leq N$) has length L_i, (2) a length range $[min\_len \cdots max\_len]$ over which consensus patterns need to be discovered, and (3) a user-specified threshold ϵ, $0\leq\epsilon\leq1$, that specifies the distance threshold beyond which no further patterns can be clustered together (a small value indicates tighter clusters and higher similarity among the discovered consensus patterns), Bio-CP returns a set of clusters; each cluster represents a group of consensus patterns for a given length within the user-specified length range. The distance between a pair of patterns can be measured using the Hamming Distance [15] or any distance metric that does not allow insertions or deletions like the substitution matrix. In this paper, we use the Hamming Distance.

Bio-CP proceeds in phases where each phase discovers the clusters for a certain length within the specified range. In the first phase (P_{min_len}), we discover the clusters for patterns of length min_len, and in the subsequent phases (P_{min_len+1}, P_{min_len+2}, $\cdots$, P_{max_len}), we incrementally extend existing patterns to discover the subsequent clusters corresponding to each length. Initially, Bio-CP divides the sequences in the database D into sliding windows of length min_len. Each sliding window W_{ij}^k is identified by three indexes i, j, and k, where i specifies the sequence identifier, j specifies the start position within sequence S_i, and k specifies the length of the sliding window. The index k is initially set to min_len and k increases by one as we move from one phase to the next one. We build a list Q that contains all possible sliding windows (Figure 1(a)). Each node in the list represents a pattern or sliding window, and consists of four fields: index i, index j, a pointer to the pattern, and a pointer to the next pattern in the list. These nodes will form the leaf level of a clustering hierarchy that will be built on top of them. In the rest of the paper, we use the terms 'window' and 'pattern' interchangeably to refer to the sliding window pattern.

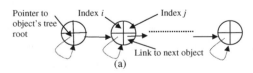

Pointer to first obj.	Pointer to second obj.	Ref_distance	Distance	Cluster_pointer
O_i	O_j	D_{i-j}	D_{i-j}	NULL
---	---	---	---	---

(b)

Fig. 1. (a) Patterns list structure (Q). (b) Sorted list structure (S).

In phase P_{min_len}, Bio-CP starts by measuring the distance between each possible pair of patterns on length min_len, and then inserting an entry for this pair into a sorted list S. List S is maintained in an ascending order of distance values such that the pairs that are the closest to each other will be at the top of the list and will be the first to be merged together. The structure of list S consists of five columns (see Figure 1(b)): The first column holds a pointer to a pattern in Q, the second column holds also a pointer to a pattern in Q, the third and fourth columns hold distance values, and if that entry will create a new node in the clustering hierarchy, then the fifth column will hold a pointer to this new node, otherwise it is NULL. The values in the *Ref_distance* column serve as reference values and will not change while proceeding from one phase to another. However, the values in the *Distance* column will be updated as we progress in the algorithm.

Our target is to generate clusters over a range of lengths. Thus, it is important to normalize the distance values relatively to the patterns lengths (the distance between any pair is always between 0 and 1.) Furthermore, generating the distance matrix as a sorted list (S) does not involve additional overhead. Indeed, since we have prior knowledge of the possible values that will be inserted into the list, (the Hamming Distance between two patterns of length L is between 0 and L) then each new entry can be directly hashed to its proper sorted location in the list. We maintain a linked list for each distance value. After all the entries are inserted into their corresponding lists, we link these lists together to form list S. Bio-CP is applied over two clustering metrics, single-link [26], presented in Section 3.1 and complete-link [18] presented in Section 3.2. Typically, single-link clustering involves less processing overhead than complete-link clustering. However, in the case of large databases, single-link clustering generates poor quality clusters due to the chaining effect [23].

3.1 Progressive Single-Link Clustering

In single-link clustering, the distance between two clusters is the distance between the closest pair of patterns in these clusters. The objective is to merge in each step the closest pair of clusters into one cluster. We refer to Bio-CP with single-link clustering by Bio-CP/S. In the first phase P_{min_len}, we scan the list S until we reach the first entry with distance larger than or equal to ϵ. Any entry after that entry has a distance value larger than or equal to ϵ. For each entry, say e, we check if the two patterns in e belong to the same cluster; If this is the case, then e is skipped, otherwise, the corresponding two clusters, say C_i and C_j, are merged into a new cluster C_k. The merge operation involves three steps: (1) Create a new cluster C_k in the clustering hierarchy. Make C_i and C_j children for C_k. (2) Traverse clusters C_i and C_j to reach all their members at the leaf level. Update the pointer associated with each member to point to cluster C_k instead of C_i or C_j. (3) Add a pointer in entry e to point to the newly created cluster C_k. Figure 2 shows an example of clustering five patterns. Note that the fourth entry in list S does not create a new cluster node, therefore its *Cluster_pointer* value remains NULL. Recall that each window is initially in

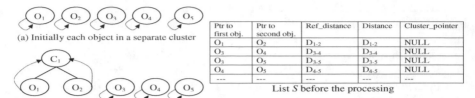

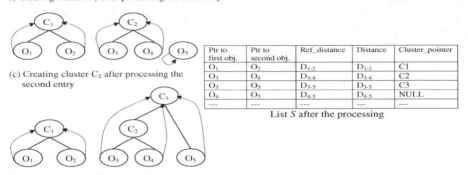

Ptr to first obj.	Ptr to second obj.	Ref_distance	Distance	Cluster_pointer
O_1	O_2	D_{1-2}	D_{1-2}	NULL
O_3	O_4	D_{3-4}	D_{3-4}	NULL
O_3	O_5	D_{3-5}	D_{3-5}	NULL
O_4	O_5	D_{4-5}	D_{4-5}	NULL
---	---	---	---	---

List S before the processing

Ptr to first obj.	Ptr to second obj.	Ref_distance	Distance	Cluster_pointer
O_1	O_2	D_{1-2}	D_{1-2}	C1
O_3	O_4	D_{3-4}	D_{3-4}	C2
O_3	O_5	D_{3-5}	D_{3-5}	C3
O_4	O_5	D_{4-5}	D_{4-5}	NULL
---	---	---	---	---

List S after the processing

(a) Initially each object in a separate cluster

b) Creating cluster C_1 after processing the first entry

(c) Creating cluster C_2 after processing the second entry

(d) Creating cluster C_3 after processing the third entry

Fig. 2. Updating the clustering hierarchy and S after processing the first four entries

a separate cluster and the pointer associated with it points to itself. Also all the pointers in the sorted list S are initially NULL. Maintaining the pointers in steps (2) and (3) is crucial for efficient processing in the subsequent phases. For example the pointer associated with each window, and maintained in step (2), allows us to detect if two members belong to the same cluster in a constant time.

Now that we generated the clusters of the consensus patterns of length min_len, we can start generating the clusters for the subsequent phases P_{min_len+1}, $P_{min\_len+2}, \cdots, P_{max\_len}$ in an incremental way. Since the windows are sorted in S based on length min_len, then all the subsequent phases will reference these windows for further extensions. In addition, these subsequent phases will reference the distance values measured in phase P_{min_len}. For this reason, each entry in S keeps a copy of this reference distance value ($Ref\_Distance$).

Generally speaking, generating the clusters in any phase P_{min_len+t} involves extending the windows in phase P_{min_len} by t letters. Our progressive processing is based on two key observations.

Observation 1. *The change in the distance values among the windows due to extending them by t letters is bound. The bound is computed using the Equation*
$$\Delta_{max\_t} = \frac{t}{min\_len+t}.$$

Observation 2. *The value of Δ_{max_t} increases monotonically with respect to t; the size of $C(t)$ is always increasing: $C(t-1) \subseteq C(t)$; where $2 \le t$.*

Observation 1 implies that the maximum change in the distance value due to appending t letters is $\pm\Delta_{max\_t}$. Therefore, the only entries in S that may cross the threshold ϵ in phase P_{min_len+t}, and thus may change the clustering hierarchy are within $\pm\Delta_{max\_t}$ from ϵ. Let $C(t)$ be the set of entries that are within $\pm\Delta_{max\_t}$ from ϵ; entries in $C(t)$ are the only entries that need to be updated in phase P_{min_len+t}. This would simply consist in comparing the newly appended t letters and modifying the distance value. Finding the entries in S that belong to $C(t)$ is performed by scanning S in both directions starting from the last processed entry in phase P_{min_len} (i.e., the last entry that has $Ref\_Distance < \epsilon$). All entries that have $Ref\_Distance$ value within $\pm\Delta_{max\_t}$ from ϵ will be in $C(t)$.

Observation 2 implies that the entries in $C(t)$ that need to be updated in phase P_{min_len+t} can be divided into two types: (1) the entries that were in $C(t-1)$ in the previous phase $P_{min\_len+t-1}$ and (2) the entries that are added during the current phase P_{min_len+t}. The entries in the latter type require comparing the newly appended t letters to update their distances. In contrast, the entries in the former type require only one letter comparison (the last appended letter) to update their distances. In summary, Observation 1 determines which entries to update and Observation 2 allows for an efficient update of these entries' distance values. The main framework for progressive processing is given in Figure 3. Note that the equation for Δ_{max_t} is general for the Hamming distance as well as for any substitution matrix having values ranging from 0 to any positive number.

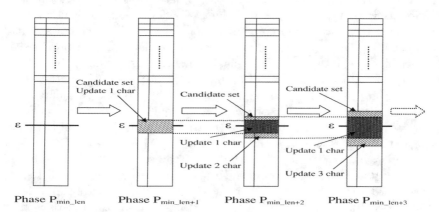

Fig. 3. Progressive processing for the candidate entries

The next step is to identify which of these entries trigger a change over the existing hierarchy; i.e., split or merge operations:

- If the distance before the update is less than ϵ and the distance after the update is still less than ϵ, then this entry will not trigger any change.
- If the distance before the update is larger than or equal to ϵ and the distance after the update is still larger than or equal to ϵ, then this entry will not trigger any change.
- If the distance before the update is less than ϵ and the distance after the update is larger than or equal to ϵ, then this entry will trigger a split to take place only if the entry has a cluster attached to it.

– If the distance before the update is larger than or equal to ϵ and the distance after the update is less than ϵ, then this entry will trigger a merge to take place only if the windows in this entry belong to different clusters.

If none of the entries in $C(t)$ triggers splits or merges, then the clusters in the current phase P_{min_len+t} will be the same as the clusters generated in the previous phase $P_{min\_len+t-1}$. If there are entries in $C(t)$ that trigger splits or merges, then the entries in $C(t)$ need to be scanned to perform the required changes.

Performing a split operation, triggered by an entry e in list S, on a cluster C involves three steps: (1) Traverse the left child cluster of C, i.e., C_1, to reach all its members in the leaf level. Update the pointer associated with each member to point to cluster C_1 instead of C. (2) Traverse the right child cluster of C, i.e. C_2, to reach all its members in the leaf level. Update the pointer associated with each member to point to cluster C_2 instead of C. (3) Invalidate the pointer attached to entry e by setting it to NULL.

Recall that the pointers maintained at the leaf members of the hierarchy and the entries in the sorted list allow performing the identification process efficiently. Detecting whether an entry in the sorted list has a cluster attached to it is performed in constant time by checking the pointer associated with this entry. Also, detecting whether two windows belong to different clusters is performed in constant time by checking the pointers associated with these windows. Finally, performing a merge or split operation is performed, as explained previously, in the order of the cluster size. The issue of maintaining the hierarchy pointers is thightly related to the union-find problem [25]. For example, another way of maintaining the pointers is to make each node points to its direct parent only. In that case, detecting whether two windows belong to different clusters or not is performed in order of the cluster size (in the worst case), however a merge or split operation is performed in a constant time.

It is important to keep the clustering hierarchy consistent while performing the splitting and merging operations. To achieve this consistency, all the split operations are performed before any merge operations in the backward direction (i.e., we process the entries of $C(t)$ in a bottom-up fashion), and then all the merge operations are performed in the forward direction (i.e., we process the entries of $C(t)$ in a top-down fashion). Algorithm 1 describes the incremental processing for discovering the clusters at phase P_{min_len+t}. The following lemma states that Bio-CP/S produces exactly the same clusters as a non-progressive single link technique. We omit the proof due to the lack of space.

Lemma 1. *Bio-CP/S produces exact results in all phases compared to the results generated by the non-progressive technique.*

3.2   Progressive Complete-Link Clustering

Bio-CP/S has two advantages: (i) it produces exact results and (ii) the single-link clustering it employs has less processing overhead than that of complete-link

---

Algorithm 1. Progressive clustering at phase P_{min_len+t}

Inputs From the Previous Phase:
  -Sorted list S
  - The Clustering hierarchy from the previous phase $P_{min\_len+t-1}$
  -Δ_{max_t-1} from the previous phase $P_{min\_len+t-1}$
Computations at the Current Phase:
  -Compute Δ_{max_t}
  -Entries within $\pm\Delta_{max\_t-1}$ from ϵ are updated
      by comparing the last letter appended to the patterns
  -Entries within $\pm\Delta_{max\_t}$ but outside $\pm\Delta_{max\_t-1}$ from ϵ
      are updated by comparing the last t letters appended to the patterns
  -Identify the entries that cross ϵ and trigger
      split or merge operations
  -Perform the required split operations in the backward
      direction (process the entries bottom-up)
  -Perform the required merge operations in the forward
      direction (process the entries top-down)

---

clustering. However, for large databases, Bio-CP/S produces clusters with poor quality due to the chaining effect. In this section, we propose Bio-CP/C where we apply Bio-CP using the complete-link metric for clustering. In complete-link clustering, the distance between two clusters is measured as the distance between the farthest pair in the two clusters. Therefore, the diameter of any resulting cluster is always less than the user specified threshold ϵ.

Building the clustering hierarchy in the first phase, P_{min_len}, is similar to building the hierarchy using the single-link metric, except in the merging condition. After constructing the sorted list S, S is scanned until we reach the first entry with a distance larger than or equal to ϵ. For each entry e in S, if the two patterns in that entry, i.e., W_1 and W_2, belong to the same cluster then e is skipped, otherwise if W_1 and W_2 belong to different clusters, i.e., C_1 and C_2, then Bio-CP/C checks whether W_1 and W_2 are the farthest pair in C_1 and C_2. If this is the case, then C_1 and C_2 are merged into a new cluster C, otherwise, entry e is skipped. The merge step in Bio-CP/C is similar to the merge step in Bio-CP/S except that we do not need to maintain the pointers attached to the entries in list S (Step 3 in the merge procedure). These pointers are used in Bio-CP/S to identify efficiently which entries will trigger splits in the subsequent phases. However, in Bio-CP/C, the splitting condition is different and does not depend on these pointers.

After generating the clusters of the consensus patterns of length min_len in phase P_{min_len}, Bio-CP/C generates the clusters for the subsequent phases P_{min_len+1}, P_{min_len+2}, $\cdots$, P_{max_len}. In each phase, i.e., P_{min_len+t}, Bio-CP/C computes $C(t)$ (the set of entries to be updated) and then updates their distance values in the same way as that of single-link clustering. Identifying which entries will trigger splits or merges is performed as follows:

- If the distance before the update is less than ϵ and the distance after the update is still less than ϵ, then this entry will not trigger any change.
- If the distance before the update is larger than or equal to ϵ and the distance after the update is still larger than or equal to ϵ, then this entry will not trigger any change.
- If the distance before the update is less than ϵ and the distance after the update is larger than or equal to ϵ, then this entry will trigger a split to take place whenever the windows in this entry belong to the same cluster.
- If the distance before the update is larger than or equal to ϵ and the distance after the update is less than ϵ, then this entry will trigger a merge to take place only if the windows in this entry belong to different clusters and are the farthest in their clusters.

The splitting condition in Bio-CP/C is less strict than that in Bio-CP/S. In Bio-CP/S, the only entry that can split an existing cluster is the one that created the cluster. However, in Bio-CP/C, any pair of windows that belong to the same cluster will split the cluster if their distance becomes larger than or equal to ϵ. This is why Bio-CP/C does not maintain pointers with the entries in the sorted list S. In addition, the merging condition in Bio-CP/C is more strict than in Bio-CP/S; a pair of windows will merge two clusters only if this pair is the farthest pair in these clusters.

Bio-CP/C and Bio-CP/S use the progressive technique in almost the same way. However, the results from Bio-CP/C are approximate in comparison with the non-progressive technique as stated in the following lemma. We omit the proof due to the lack of space.

Lemma 2. *Bio-CP/C produces approximate results in comparison to the results produced by the non-progressive technique. However, the generated clusters still satisfy the condition that the diameter of any generated cluster is less than ϵ.*

3.3   Post-processing Phase

The post-processing phase allows to apply any available domain-specific requirements to refine the discovered consensus patterns. Here are some examples of biological requirements that can be applied for discovering motifs and tandem repeat. (a) The user may specify a minimum size for the desired clusters, such that any cluster that contains fewer patterns than the specified minimum size will be ignored. (b) The user may specify any position requirements for the discovered patterns. For example, whether the desired tandem repeats should appear immediately after each other or a gap is allowed between the repeats. (c) A DNA palindrome [8] is a sequence whose inverse complement is the same as the original sequence. The user may specify that the desired motif patterns should have at least a certain palindrome degree. In this case, we check each pattern against the specified palindrome threshold and we qualify only the patterns with higher palindrome degree. (d) The user may specify whether the desired consensus patterns can overlap or not. For example, the occurrences of real motifs usually do not overlap each other. In biological databases, although motifs and tandem repeats have different post-processing requirements, Bio-CP allows both types of patterns to be discovered in a single run.

4   Scalability Issues in Bio-CP

While Bio-CP shows significant improvement compared to non-progressive clustering techniques, it still involves a high overhead in the first phase due to computing and storing the distance matrix. In this section, we propose a top-k nearest-neighbor method to reduce the storage overhead and a heuristic to significantly reduce the number of comparisons needed to get the top-k nearest-neighbors for each pattern. We call the new algorithm Bio-CP/K. Another issue with Bio-CP is that at some point the overhead to process and maintain the entries in $C(t)$ may become comparable to the overhead of resorting the entries in list S. Bio-CP may need to reset the computations for list S. Due to space limitations, we do not discuss the possible solutions for this issue.

4.1   Storage Reduction Using Top-k Nearest-Neighbor

The top-k nearest-neighbor method is well known to reduce the size of the distance matrix [12]. In this method, only the top-k nearest-neighbors for each pattern are stored and clustering techniques are applied over these stored patterns only. While suing the top-k nearest-neighbor method with hierarchical clustering techniques is straightforward, using it with Bio-CP is nontrivial. Due to progressive processing, the lengths of the patterns increase from one phase to the next. Hence, the top-k nearest-neighbors for each pattern may change across phases. Bio-CP/K ensures that, in each phase, every pattern will have its top-k nearest-neighbor patterns among the patterns being processed.

Since patterns expand across the phases, Bio-CP/K stores for each pattern the union of the pattern's top-k nearest-neighbors over the different phases. For example, if the set of top-k nearest-neighbors for pattern x in phase P_{min_len+t} is K_t, then Bio-CP/K stores for x the union set $\cup_x = \cup(K_t)$; where $0 \le t \le max\_len - min\_len$. In this case, while x expands over the different phases, Bio-CP/K ensures that x's top-k nearest-neighbors are among the patterns being processed. Bio-CP/K computes the nearest-neighbors for each pattern while constructing the sorted list S and before any of the processing phases. Constructing the union of the top-k nearest-neighbors for pattern x is performed incrementally as follows:

1. Initially $\cup_x$ is empty

2. Compare pattern x with the remaining patterns based on length min_len. The result from the comparison is a list L sorted based on the distance values.

3. Add the first k patterns in L to $\cup_x$. Let the distance value of the last pattern of these k patterns be K_Dist.

4. To expand pattern x and re-compute x's top-k nearest-neighbors, we repeat the following steps for t from 1 to $(max\_len - min\_len)$:
   - The maximum change in the distance value due to appending t letters is $\Delta_{max\_t} = \frac{t}{min\_len+t}$
   - Check the patterns in L that are within $\pm 2\Delta_{max\_t}$ from K_Dist by appending and comparing t letters to the patterns with pattern x. These patterns are the only ones that may substitute each other as the top-k nearest-neighbors.
   - If any pattern of the new top-k nearest-neighbors is not in $\cup_x$, we add it to $\cup_x$.

The reason for considering $\pm 2\Delta_{max\_t}$ instead of $\pm\Delta_{max\_t}$ as in Bio-CP is that we do not care about the absolute value of K_Dist. Instead, we care about the relative order among the patterns around the K_Dist point.

After computing x's nearest-neighbors $\cup_x$, Bio-CP/K inserts pairs (x, y) $\forall y \in \cup_x$ in the sorted list S. Then the progressive processing is applied over S as discussed in Section 3. Comparing Bio-CP/K to the non-incremental technique, we observe the following. First, Bio-CP/K involves some overhead in constructing $\cup_x$. However, the non-incremental technique has to perform steps (2) and (3) independently for each pattern in each length, which clearly involves a much higher overhead. Second, since each pattern will have at least its top-k nearest-neighbors in S in each phase, then the results from Bio-CP/K are at least as good as the non-incremental technique.

4.2   Processing Time Reduction

In this section, we propose a heuristic to reduce the average number of comparisons needed to find the nearest neighbors for each pattern in Bio-CP/K. The main idea is that from comparing a given pattern x with the other patterns in the database, we can obtain the top-k nearest-neighbors for several patterns other than x in an efficient and less expensive way.

When pattern x is compared with other patterns in the database and list L is constructed (see Section 4.1), the list is logically partitioned into groups G_0, G_1, ..., G_{min_len}; where group G_i contains the patterns that have i mismatches with pattern x. It is clear that we can directly get the nearest neighbors for all the patterns that belong to group G_0. They will have the same nearest neighbors as pattern x. Similarly, for any pattern y in group G_1, y's nearest neighbors can be obtained efficiently because we know to a large extent in which groups these nearest neighbors will be. For example, the number of mismatches between pattern y and any pattern in group G_0 is one, the number of mismatches between pattern y and any pattern in group G_1 is either zero, one or two, and the number of mismatches between pattern y and any pattern in group G_2 is either one, two or three, and so on. Therefore, the comparisons can be performed incrementally based on the need for more patterns to be added to the nearest neighbor list. In that case we avoid many unnecessary comparisons. We should note that the patterns that match exactly with pattern y exist only in group G_1 and these patterns will have the same nearest neighbor list as y. The same idea applies for the other groups. However, the power of the heuristic relies on processing only the first few groups since the uncertainty in the number of mismatches is very small for these groups. Most of the comparisons will lead to patterns in the nearest neighbor list. Using the proposed heuristic allows one scan to the database to generate the required nearest neighbors for several patterns. As a result, the overall number of comparisons is reduced significantly.

5   Experimental Results

We evaluate the performance of Bio-CP for discovering motifs and tandem repeats using real datasets from the E.coli genome sequence. We consider two measures of performance; the processing time and the cluster validity. For the latter, we use *Jaccard* and *Rand coefficients* [14] to measure the similarity among the clusters generated from Bio-CP and the non-progressive clustering techniques. All measures for the non-progressive techniques are cumulative values to generate the desired clusters over multiple lengths. The non-progressive single-link

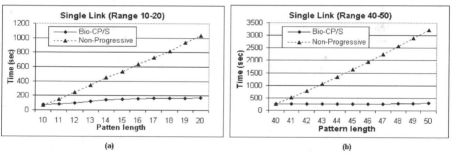

Fig. 4. Processing time of Bio-CP/S and the non-progressive technique

technique simulates the MOPAC algorithm [13]. In the first experiment, we measure the performance of Bio-CP/S. A file of size 25,000 bases is used, ϵ is assigned a value of 0.3, and two length ranges are evaluated. Figure 4 illustrates that Bio-CP/S and the non-progressive technique take almost the same time in the first phase in which the hierarchy is generated. However, in the subsequent phases, Bio-CP/S takes much less time than the non-progressive technique that rebuilds the hierarchy from scratch. Figure 4 shows that the processing time for building the hierarchy in the first phase in the case of the range [40...50] is higher than in the case of the range [10...20]. However, the time needed to progress from one phase to the next one in the case of the range [40...50] is less than in the case of the range [10...20]. The reason being that the effect of extending longer patterns is less than the effect of applying the same extension over shorter patterns. The clusters generated from both techniques in this experiment are exactly the same since we are using the single-link metric.

In Figure 5, we present the results of applying Bio-CP/C over the same file of size 25,000. The ϵ threshold is assigned a value of 0.5. The behavior of Bio-CP/C

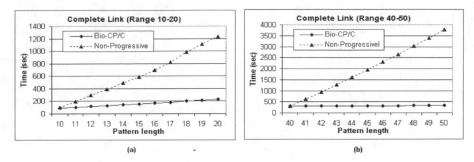

Fig. 5. Processing time of Bio-CP/C and the non-progressive technique

Length	Jac. coeff	Rand coeff	Length	Jac. coeff	Rand coeff	Length	Jac. coeff	Rand coeff
10	1	1	40	1	1	10	0.9364	0.999989
11	0.884464	0.999987	41	0.917552	0.999991	11	0.873205	0.999985
12	0.841966	0.999955	42	0.832075	0.999985	12	0.848723	0.999983
13	0.808337	0.999941	43	0.835635	0.999985	13	0.841132	0.999983
14	0.833241	0.999951	44	0.819972	0.999974	14	0.807214	0.999977
15	0.824595	0.999948	45	0.830797	0.999982	15	0.815188	0.999982
16	0.793678	0.999923	46	0.802416	0.999972	16	0.780025	0.999968
17	0.78032	0.999923	47	0.807746	0.999972	17	0.81346	0.999982
18	0.802123	0.999932	48	0.824261	0.999978	18	0.768912	0.999959
19	0.781047	0.999923	49	0.785684	0.999969	19	0.74306	0.99994
20	0.75642	0.999902	50	0.788491	0.99997	20	0.719853	0.999926
(a)			(b)			(c)		

Fig. 6. Clustering validation degree

and non-progressive techniques is very similar to that in the case of Bio-CP/S. However, the generated clusters from Bio-CP/C in this experiment are approximate clusters. Figures 6(a) and 6(b) give the validity degree of the clusters generated in the case of the ranges [10...20] and [40...50], respectively. The figure illustrates that the coefficient values are very close to 1, which means that the clusters generated from incremental processing are very similar to the ones generated from the non-progressive technique. In addition, the figure illustrates that the *Rand* coefficient detects higher similarity degree among the clusters than the *Jaccard* coefficient. The reason for this difference is that the *Jaccard* coefficient does not take into account one similarity factor, namely the number of pairs, say (w_1, w_2), for which both clustering techniques assign w_1 and w_2 patterns to different clusters. The *Rand* coefficient takes this factor into account.

In Figure 7, we present the results of applying Bio-CP/KC, (Bio-CP using complete-link metric and the top-k nearest-neighbors method). We use a file of size 150,000 bases, with ϵ assigned to 0.5. In this experiment, we store the top 0.1% nearest-neighbors for each pattern. Figure 7(a) gives the processing overhead of the various techniques. The figure illustrates that the non-progressive technique involves infeasible processing overhead in the case of relatively large files. In Figure 7(b), we present the effect of applying Bio-CP/KC along with the proposed heuristic in computing the nearest-neighbors when compared to the naive way. The figure illustrates significant reduction in the processing time due to reducing the number of comparisons. The validity measure for the clusters

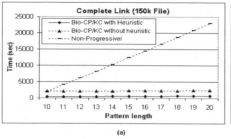
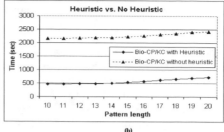

(a)   (b)

Fig. 7. (a) Processing time of Bio-CP/KC and the non-progressive technique. (b) Heuristic versus naive method.

generated in this experiment is given in Figure 6(c). Note that the coefficient values for the patterns of length 10 are no longer equal to 1. This slight dissimilarity in the first phase is due to the difference between the nearest-neighbor sets maintained by Bio-CP/KC and the non-progressive techniques.

We run several experiments to compare Bio-CP/C with the MEME algorithm [6]. The closeness of the discovered motifs is highly affected by the ϵ threshold. With ϵ assigned to 0.3 or 0.4, Bio-CP/C usually splits the motifs discovered by MEME into multiple motifs. However, with ϵ assigned to 0.6, the motifs discovered by MEME are a subset of the motifs discovered by Bio-CP/C. This indicates that 0.6 is a reasonable value for ϵ. Bio-CP/C always produces more candidate motifs than MEME. However, MEME annotates each motif with more information such as the E-value and background probabilities due its statistical nature.

6   Conclusions

In this paper, we proposed Bio-CP, a progressive hierarchical clustering technique for discovering consensus patterns, namely motifs and tandem repeats, in biological databases over a range of possible lengths. The progressive nature of the hierarchical clustering algorithm makes it scalable and efficient. Bio-CP is also applicable to a wide range of applications since any domain-specific requirements are applied in a post-processing phase. We also proposed several scalability techniques to enhance the performance of Bio-CP in terms of processing time and storage. Our experiments illustrated that Bio-CP scales very well with respect to the processing time, and the clustering validation degrees. In particular, Bio-CP has more than 500% processing time improvement.

References

1. C. C. Aggarwal. On effective classification of strings with wavelets. In *Proceedings of the 8th ACM SIGKDD*, 163–172, 2002.
2. B. Goethals. Survey on frequent pattern mining. *Manuscript*, 2003.
3. R. Agrawal, C. Faloutsos, and A. N. Swami. Efficient similarity search in sequence databases. In *FODO*, 69–84, 1993.

4. R. Agrawal and R. Srikant. Fast algorithms for mining association rules. In *VLDB*, 487–499, 1994.
5. W. G. Aref and D. Barbara. Supporting electronic ink databases. In *Information Systems: An International Journal 24(4)*, 303–326, 1999.
6. T. Bailey, C. Elkan, and B. Grundy. The meme system: Multiple EM for motif elicitation, http://bioweb.pasteur.fr/seqanal/motif/meme/.
7. T. L. Bailey and C. Elkan. Fitting a mixture model by expectation maximization to discover motifs in biopolymers. In *ISMB*, 28–36, 1994.
8. T. L. Bailey and C. Elkan. The value of prior knowledge in discovering motifs with meme. In *ISMB*, 21–29, 1995.
9. G. Benson. Tandem repeats finder: a program to analyze dna sequences. In *Nucleic Acids Research*, volume 27, 573–580, 1999.
10. P. Berkhin. Survey of clustering data mining techniques. San Jose, CA, 2002.
11. J. Buhler and M. Tompa. Finding motifs using random projections. In *RECOMB*, 69–76, 2001.
12. E. Fix and J. L. Hodges. Discriminatory analysis, nonparametric discrimination: Consistency properties. In *USAF School of Aviation Medicine, Project 21-49004, Report 4*, 1951.
13. R. Ganesh, T.R. Ioerger, and D.A. Siegele. Mopac: Motif finding by preprocessing and agglomerative clustering from microarrays. In *PSB*, 41–52, 2003.
14. M. Halkidi, Y. Batistakis, and M. Vazirgiannis. Clustering algorithms and validity measures. In *Tutorial paper, SSDBM*, 2001.
15. R. W. Hamming. Coding and information theory. In *Prentice-Hall*, 1980.
16. G.Z. Hertz and G.D. Stormo. Identifying dna and protein patterns with statistically significant alignments of multiple sequences. In *Bioinformatics*, 15:563–577, 1999.
17. H. V. Jagadish, N. Koudas, and D. Srivastava. On effective multi-dimensional indexing for strings. In *SIGMOD*, 403–414, 2000.
18. B. King. Step-wise clustering procedures. In *J. Am. Stat. Assoc. 69,*, 1967.
19. G. M. Landau and J. P. Schmidt. An algorithm for approximate tandem repeats. In *CPM*, 120–133, 1993.
20. C.E. Lawrence, S.F. Altschul, M.S. Boguski, J.S. Liu, A.F. Neuwald, and J.C. Wootton. Detecting subtle sequence signals a gibb's sampling strategy for multiple alignment. In *Science*, 262:208–214, 1993.
21. C. Liang. Copia: A new software for finding consensus patterns in unaligned protein sequences. In *Master thesis, University of Waterloo*, 2001.
22. G. Myers and M. Sagot. Identifying satellites and periodic repetitions in biological sequences. In *Journal of Computational Biology*, volume 10, 10–20, 1998.
23. G. Nagy. State of the art in pattern recognition. In *Proc. IEEE 56*, 1968.
24. P.A. Pevzner and S. Sze. Combinatorial approaches to finding subtle signals in dna sequences. In *ISMB*, 269–278, 2000.
25. J. A. La Poutr;. New techniques for the union-find problem. In *SIAM*, 54–63, 1990.
26. P. H. Sneath and R. R. Sokal. Numerical taxonomy. *Freeman,London,UK.*,1973.
27. W. B. Frakes and R. B. Yates, editors. *Information Retrieval: Data Structures and Algorithms*. Prentice-Hall, 1992.
28. E. M. McCreight. A space-economical suffix tree construction algorithm. *Journal of ACM*, 23(2):262–272, 1976.
29. D. Gusfield. *Algorithms on strings, trees, and sequences: computer science and computational biology*. Cambridge University Press, New York, NY, USA, 1997.
30. H. Hamada, M. Seidman, B. Howard, and C. Gorman. *Enhanced gene expression by the poly(dT-dG) poly(dC-dA) sequence* Molecular and Cellular Biology, 1984.

Comparison of Modularization Methods in Application to Different Biological Networks

Zhuo Wang[1,4], Xin-Guang Zhu[2], Yazhu Chen[1], Yixue Li[4], and Lei Liu[3,4]

[1] Biomedical Instrument Institute, Shanghai Jiao Tong University, 1954 Huashan Rd, Shanghai, 200030, China
[2] Department of Plant Biology, University of Illinois at Urbana-Champaign, 1201 W. Gregory Dr., Urbana, Illinois 61801, USA
[3] The W. M. Keck Center for Comparative and Functional Genomics, University of Illinois at Urbana-Champaign, 1201 W. Gregory Dr., Urbana, Illinois 61801, USA
[4] Shanghai Center for Bioinformation Technology, 100 Qinzhou Rd, 12th Floor, Shanghai, 200235, China

Abstract. Most biological networks have been proposed to possess modular organization, which increases the robustness, flexibility, and stability of networks. Many clustering methods have been used in mining biological data and partitioning complex networks into functional modules. Most of these methods require presetting the number of modules and therefore can potentially obtain biased results. The Markov clustering method (MCL) and the simulated annealing module-detection method (SA) eliminate this requirement and can objectively separate relatively dense subgraphs. In this paper, we compared these two module-detection methods for three types of biological data: protein family classification, microarray clustering, and modularity of metabolic networks. We found that these two methods show differential advantages for different biological networks. In the case of the gene network based on Affymetrix microarray spike data, MCL exactly identified the same number of groups and same contents in each group set by the spike data. In the case of the gene network derived from actual expression data, although neither of the two methods can perfectly recover the natural classification, MCL performs slightly better than SA. However, with increased random noise added to the gene expression values, SA generates better modular structures with higher modularity. Next we compared the modularization results of MCL and SA for protein family classification and found the modules detected by SA could not be well matched with the Structural Classification of Proteins (SCOP database), which suggests that MCL is ideally suited to the rapid and accurate detection of protein families. In addition, we used both methods to detect modules in the metabolic network of *E. coli*. MCL gives a trivial clustering, which generates biologically insignificant modules. In contrast, SA detects modules well corresponding to the KEGG functional classification. Moreover the modularity for several other metabolic networks detected by SA is also much higher than that by MCL. In summary, MCL is more suited to modularize relatively complete and definite data, such as a protein family network. In contrast, SA is less sensitive to noise such as experimental error or incomplete data and outperforms MCL when modularizing gene networks based on microarray data and large scale metabolic networks constructed from incomplete databases.

Keywords: Markov clustering (MCL), simulated annealing (SA), module, network.

M.M. Dalkilic, S. Kim, and J. Yang (Eds.): VDMB 2006, LNBI 4316, pp. 185–195, 2006.
© Springer-Verlag Berlin Heidelberg 2006

1 Introduction

Many biological systems such as gene regulatory networks, protein-protein interaction networks, metabolic networks, and ecological networks are complex network. Metabolic networks and the network of protein interactions exhibited modularized structure or functional modules [1-5]. Tanay *et al.* performed analysis of a highly diverse collection of genome-wide data sets, including gene expression, protein interactions, transcription factor binding, and revealed modular organization at all these different levels [6]. Therefore, modularized organization, a division of network nodes into groups within which the network connections are dense, but between which they are sparse, is a common property of complex biological networks. Modular structures have been suggested to be associated with increased robustness, flexibility, and stability of the underlying network [7-10].

Given such universal existence of modular structure and potential functional significance, much effort has been devoted to studying the decomposition of complex biological networks into modules. The problem of identifying modules in a complex network is closely related to the graph partitioning problem, i.e. the separation of sparsely connected dense subgraphs from each other. Several graph clustering techniques have been applied to network module detection such as Markov clustering (MCL), iterative conductance cutting, and geometric minimal spanning tree clustering.

The idea behind Markov clustering is to simulate random walks or flow within the network, strengthening flow where the flow rate is high and weakening flow where the flow rate is low [11]. The key intuition is that a random walk that visits a dense cluster will likely keep in the cluster until most of its vertices have been visited. Flow simulation can be done by computing powers of a suitable Markov matrix. The first operation used by MCL is expansion through matrix multiplication, which simulates the spreading of free flow. The second is inflation, which is mathematically speaking a Hadamard power followed by a diagonal scaling. Inflation models the contraction of flow, it becoming thicker in regions of higher current and thinner in regions of lower current. The MCL process causes flow to spread out within natural clusters and evaporate between clusters [11].

The basis of iterative conductance cutting is to iteratively split clusters using minimum conductance cuts [12]. Finding a cut with minimum conductance is NP-hard, therefore a poly-logarithmic approximation algorithm is used. Geometric minimal spanning tree clustering is a graph clustering algorithm combining spectral partitioning with a geometric clustering technique [13]. The experimental study confirms that MCL performs better in many cases, although it is slower than iterative conductance cutting and geometric minimal spanning tree clustering [14].

Guimerà and Amaral identified modules in metabolic network by maximizing the network's modularity using simulated annealing (SA) [15, 16]. At each temperature, the network structure was updated by two kinds of random movements: individual node movements from one module to another and collective movements that involve either merging two modules or splitting a module. The optimal module partition can be obtained until the modularity value reach to the maximal. By relating the metabolites in any given module to nine KEGG major pathways, they validated that more than one-third of the metabolites in any module belong to a single pathway, which can provide a functional cartographic representation of the complex network [15].

Two of the modularization methods mentioned above and many others such as k-Means and k-Medoid require presetting the number of modules and therefore can potentially obtain biased results [17, 18]. MCL and SA eliminate this requirement and can objectively separate relatively dense subgraphs [11, 15, 16]. Which clustering algorithm performs better? In this paper, we compared two modularization methods, MCL and SA, using different biological networks such as gene networks reconstructed from microarrays, protein family networks, and metabolic networks. We hypothesize that MCL and SA show differential advantages for different network data due to the different underlying principle of these two algorithms.

2 Methods

2.1 Evaluation on the Modular Structure of Complex Networks

According to Guimerà [19], for a given partitioning of the nodes of a network into modules, the modularity M of this partition is defined as:

$$M = \sum_{s=1}^{r} \left[\frac{l_s}{L} - \left(\frac{d_s}{2L} \right)^2 \right]$$

where r is the number of modules, L is the number of links in the network, ls is the number of links between nodes in module s, and ds is the sum of the degrees of the nodes in module s. This definition of modularity implies that $M<=1$, and that $M=0$ for a random partition of the nodes.

MCL algorithm can be downloaded from MCL webpage (http://micans.org/mcl) and was applied here with an inflation parameter typically set to $I = 2.0$ and with other parameters at default values. The program for SA algorithm was derived from Guimerà and Amaral. We set the iteration factor to $f=1$ and cooling factor to $c=0.99$.

2.2 Construction and Modularization of Gene Network from Microarray Data

We used a set of microarray data derived from Affymetrix Human Gene Chips U133 as an example. The data are downloaded from Affymetrix website [20]. This data set consists of 3 technical replicates of 14 separate hybridizations of 42 spiked transcripts in a complex human background at concentrations ranging from 0.125pM to 512pM. Thirty of the spikes are isolated from a human cell line, four spikes are bacterial controls, and eight spikes are artificially engineered sequences believed to be unique in the human genome. These 42 spikes were initially selected based on their absent expression in background total RNA isolated from a HeLa cell line (ATCC CCL-13) by both GeneChip and Taqman assays, which were listed in Table 1.

Table 1. The 42 spikes of Affymetrix Human Gene Chips U133

Group ID	Gene ID
1	203508_at 204563_at 204513_s_at
2	204205_at 204959_at 207655_s_at
3	204836_at 205291_at 209795_at
4	207777_s_at 204912_at 205569_at
5	207160_at 205692_s_at 212827_at
6	209606_at 205267_at 204417_at
7	205398_s_at 209734_at 209354_at
8	206060_s_at 205790_at 200665_s_at
9	207641_at 207540_s_at 204430_s_at
10	203471_s_at 204951_at 207968_s_at
11	AFFX-r2-TagA_at AFFX-r2-TagB_at AFFX-r2-TagC_at
12	AFFX-r2-TagD_at AFFX-r2-TagE_at AFFX-r2-TagF_at
13	AFFX-r2-TagG_at AFFX-r2-TagH_at AFFX-DapX-3_at
14	AFFX-LysX-3_at AFFX-PheX-3_at AFFX-ThrX-3_at

We calculated the Pearson correlation coefficients among these genes based on the relationship of their expression value. Then a gene network can be constructed, where nodes represent genes and there will be an edge connecting two genes if the correlation coefficient between them is higher than a threshold. Accordingly, we constructed a small network involving these 42 genes using their spikes data, which has the exact same structure with 42 links when the correlation coefficient threshold is either 0.5 or 0.8. In addition, we also constructed the network including these 42 genes based on their actual expression data from the microarray. This network has 173 edges and 42 edges respectively under the condition that the correlation coefficient cutoff is 0.5 or 0.8. We choose the first one for further analysis.

2.3 Protein Family Detection

The MCL algorithm has been successfully applied in the field of protein family detection [21, 22]. Harlow *et al.* collected 13,764 proteins from the Protein Data Bank (PDB) and calculated the similarity between each protein pair based on all-versus-all BLASTp. A protein connection network was then constructed, where the nodes represented proteins and edges represented sequence similarity between proteins. Using MCL they divided this protein network into many clusters which are well matched with the protein families in the Structural Classification of Proteins (SCOP database) [22]. Here we used their protein dataset of PDB and modularized it by the SA algorithm, in order to compare with MCL.

2.4 Metabolic Network Reconstruction and Module Detection

There has been no study using MCL to detect modules in metabolic networks, Guimerà and Amaral used SA to find modules in the metabolic network of *E. coli* [15, 16]. In this study, we used both MCL and SA to partition the metabolic network of *E. coli*, which is widely accepted as a benchmark with complete annotation. Different from Guimerà and Amaral, who constructed the metabolic network using compounds as nodes, we constructed an enzyme-centric graph to represent the metabolic network, where vertices represent enzymes and edges represent compounds. A directed edge from enzyme E_1 to enzyme E_2 exists if and only if E_1 catalyzes a reaction generating a product A which is used as a substrate in the reaction catalyzed by E_2. Reversible reactions are considered as two separate reactions. We extracted the pathway information of *E. coli* from KEGG database, consisting of 587 enzymes, 1163 metabolites, and 1810 reactions, thus resulted a graph with 587 nodes and 7197 edges.

3 Results

3.1 SA Outperforms MCL with the Increasing Noise in the Microarray Data

The 42 spikes network is a benchmark, which has 14 groups and each including 3 genes. Such veritable classification can be accurately recovered by MCL with high modularity value of 0.9286. However, SA could not give a reasonable result because it disregards the isolated nodes. For the network based on the actual expression data of 42 genes, MCL and SA detected 4 and 5 modules respectively, with modularities of 0.6112 and 0.5351 respectively. Then we tested the possible influence of inaccurate microarray data on modularization results by randomly adding Gaussian noise to the original expression value. The changes of modularization results by MCL and SA with the increasing noise are shown in Table 2.

Table 2. Modularization results of 42 spikes network by MCL and SA at different noise level

	Correlation coefficient threshold	Number of links	Number of modules MCL/SA	Modularity value MCL/SA
Origin	0.5	173	4 / 5	0.6112 / 0.5351
5 db Gaussian noise	0.5	142	5 / 4	0.5971 / 0.6127
10 db Gaussian noise	0.5	110	6 / 5	0.5573 / 0.6174
20 db Gaussian noise	0.5	24	19 / 4	0.4740 / 0.6191

MCL generated significantly better modularity than SA for the original network. With increased noise, SA outperforms MCL. In addition, the number of modules detected by SA does not change as much as MCL, which demonstrates that SA is less sensitive to noise than MCL.

3.2 MCL Generates More Reasonable Classification for Protein Family

For the 13,764 proteins dataset collected from PDB, Harlow *et al.* divided them into 3503 modules by MCL with an inflation parameter of 2.0. According to the within-links and between-links of the MCL modules, we computed the modularity value as 0.7255. They used SCOP families (clusters) to interpret the clustering of PDB sequences, using only 6435 proteins among the 13,764 that have exactly one PDB ID corresponding to one sequence entry, and are annotated with exactly one SCOP family each to ensure one-to-one mappings between PDB entries, PDB IDs, and SCOP families. They found that most of the clusters that include at least two nodes are pure by SCOP family; 91% of these pure clusters contain only one single SCOP family. Remarkably, the proteins in most MCL clusters belong to the same SCOP family, which indicates that MCL can classify proteins rather well according to their family classifications.

We used SA to modularize the same network and found only 12 modules and a modularity of 0.5070. Since SA disregards the isolated nodes when modularizing the whole network, many proteins disconnected from others can not be assigned into any clusters. The 12 modules detected by SA include only 1069 proteins, among which only 198 proteins having exactly one SCOP family. A large proportion of the 6435 one-to-one mapping proteins are not involved in the SA clusters, therefore it is not meaningful to assess the SA result and we conclude that SA is not suitable for protein family classification.

3.3   The Modules Detected By SA in Metabolic Network Are More Biological Meaningful

MCL decomposed *E. coli* enzyme network into 45 modules. The sizes of modules exhibit a power-law distribution, where one or two large modules include many enzymes from several unrelated biological pathways, and 28 modules contain no more than four enzymes. In contrast, SA detected 10 modules in this enzyme network and the sizes of modules are relatively uniform. The detailed list of enzymes in each module by both MCL and SA can be seen in the supplementary findings. The distributions of module sizes by these two methods are shown in Figure 1.

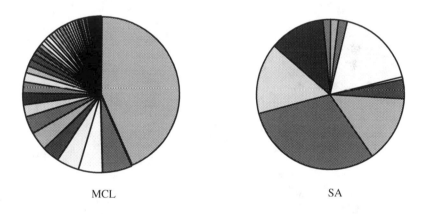

MCL   SA

Fig. 1. The distribution of module size in E. coli metabolic network by MCL and SA

It is remarkable that MCL generated more small modules, which are difficult to map to higher level functional categories. According to Guimerà and Amaral [15, 16], we mapped the modules detected by SA to KEGG functional classifications, which include nine major pathways: carbohydrate metabolism, energy metabolism, lipid metabolism, nucleotide metabolism, amino-acid metabolism, glycan biosynthesis and metabolism, metabolism of cofactors and vitamins, biosynthesis of secondary metabolites, and biodegradation of xenobiotics. As shown in Figure 2, almost every module assigns more than 30% of enzymes to one specific pathway. All of the enzymes in module 1 correspond to carbohydrate metabolism. Several modules consist of enzymes mainly involved in amino acid metabolism.

The modularity value of the modularization of the *E. coli* metabolic network by MCL and SA are 0.3215 and 0.6012 respectively, indicating the partitioning by SA is more modular and reliable. In our previous study, we have compared the module structures of metabolic networks in chloroplasts, several photosynthetic bacteria, *Cyanidioschyzon merolae*, and *Arabidopsis thaliana* using both SA and MCL. The modularization results and modularity values are shown in Table 3. All modularity values computed by SA are substantially higher than that by MCL. Another interesting finding is that the modularity values by SA for all these species are very similar, but the values computed by MCL fluctuate widely among different species. This suggests that MCL is more sensitive to network changes. These results strongly demonstrate that SA highly outperforms MCL to mine the functional modules in metabolic networks.

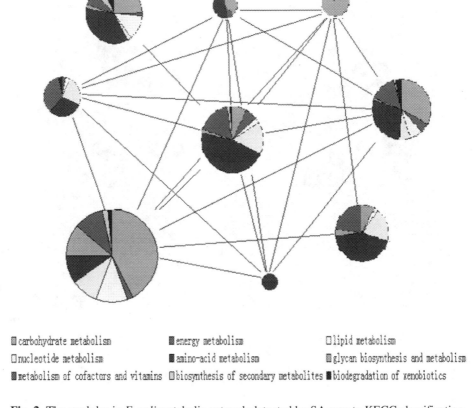

■ carbohydrate metabolism   ■ energy metabolism   □ lipid metabolism
□ nucleotide metabolism   ■ amino-acid metabolism   ■ glycan biosynthesis and metabolism
■ metabolism of cofactors and vitamins   □ biosynthesis of secondary metabolites   ■ biodegradation of xenobiotics

Fig. 2. The modules in *E. coli* metabolic network detected by SA map to KEGG classification

Each circle represents a module and is coloured according to the KEGG pathway classification of the enzymes it contains.

Table 3. The module numbers and modularity values of several metabolic networks analyzed by SA and MCL

Species	Number of linkage	SA		MCL	
		Number of modules	Modularity	Number of modules	Modularity
Chloroplast	2777	12	0.6292	48	0.5716
Ana	3761	9	0.5975	39	0.4959
Cte	2770	7	0.6114	32	0.3092
Gvi	3194	8	0.6193	34	0.5115
Pma	2910	9	0.6072	31	0.2153
Pmm	2829	9	0.6263	31	0.2409
Pmt	2883	9	0.6257	31	0.5398
Syn	3429	7	0.6231	38	0.4906
Syw	3267	8	0.6238	30	0.2512
Tel	2953	8	0.6241	40	0.5105
E.coli	7197	10	0.6012	45	0.3215
Ath	5188	11	0.5997	33	0.1962
Cme	3265	9	0.6059	34	0.4718

4 Discussion

Besides MCL and SA, some usual clustering methods can be theoretically used for network partitioning, such as hierarchical clustering, k-means and k-Medoid. The result of hierarchical clustering is represented as a tree-like shape, and the cluster is produced by cutting edges of the tree with a threshold, which is difficult to be given [17]. For k-Means and k-Medoid, the partition size is required to be specified in advance [18]. However for complex biological networks, it is very difficult and impractical to preset the module size. Predominantly, both MCL and SA methods need not specify the number of modules and cutting threshold. Clustering of microarray data has been widely studied and many effective methods have been developed, but few focus on graph clustering to solve this problem. Here we found MCL can perfectly cluster the benchmark gene network comprising 42 gene spikes, which indicates MCL can be used for clustering very high-quality microarray data. With increasing experimental noise, SA produces better results with higher modularity than MCL, implying SA is less sensitive to noise. In fact, a majority of microarray data inevitably suffers from contamination, so for gene networks constructed from microarray data, SA can give relatively reliable partitioning and find related gene groups.

MCL is ideally suited to the rapid and accurate detection of protein families, which has been rigorously tested and validated on a number of very large databases, including PDB, SwissProt, InterPro, and SCOP [21, 22]. We found the modules detected by SA in protein similarity networks could not match well with the families in the SCOP database, which again suggests MCL works best for protein family classification. Usually, a protein network based on sequence similarity is well annotated, rather definite, and unlikely to be affected by experiment or environment. Consequently MCL is more suited than SA to modularize relatively complete and definite networks.

For modularization of metabolic networks, MCL often gives a trivial clustering, which may generate biologically insignificant modules. In contrast, SA detected moderately sized modules of *E. coli* metabolic network, which correspond well to the KEGG functional classifications. In addition, the modularity by SA for chloroplasts, several photosynthetic bacteria, *Cyanidioschyzon merolae*, and *Arabidopsis thaliana* are all much higher than the modularity by MCL. It is well known that this metabolic pathway information is derived from metabolic databases, so there are inevitably many incomplete enzymes and reactions, especially for newly sequenced organisms. Because SA is less sensitive to noise such as incomplete data, it can produce more meaningful modular structures of whole genome-level networks than MCL. Moreover, the great fluctuations of modularity values among different metabolic networks by MCL confirm it is more sensitive to noise than SA.

In conclusion, SA is less sensitive to noise such as experimental error or incomplete data, so it outperforms MCL when modularizing gene networks based on microarray data and large scale metabolic networks constructed from databases. In contrast, MCL is more suitable for relatively complete and definite data, so it can accurately and rapidly detect functional families in protein networks.

Acknowledgement

We would like to thank Dr. Jenny Drnevich for providing the microarray dataset and reviewing of the manuscript. This work was supported by grants from the National "973" Basic Research Program of China (2004CB518606), and the Fundamental Research Program of Shanghai Municipal Commission of Science and Technology (04DZ14003).

References

1. Hartwell, L.H., Hopfield, J.J., Leibler, S., Murray, A.W.: From molecular to modular cell biology. Nature, Vol. 402 (1999) 47–52
2. Ravasz, E., Somera, A.L., Mongru, D.A., Oltvai, Z.N., Barabási, A.L.: Hierarchical organization of modularity in metabolic networks. Science, Vol. 297 (2002) 1551–1555

3. Rives, A.W., Galitski, T.: Modular organization of cellular networks. Proc. Natl. Acad. Sci., Vol. 100. U. S. A. (2003) 1128–1133
4. Spirin, V., Mirny, L.A.: Protein complexes and functional modules in molecular networks. Proc. Natl. Acad. Sci., Vol. 100. U. S. A. (2003) 12123–12128
5. Wilhelm, T., Nasheuer, H.P., Huang, S.: Physical and Functional Modularity of the Protein Network in Yeast. Molecular & Cellular Proteomics, Vol. 2 (2003) 292-298
6. Tanay, A., Sharan, R., Kupiec, M., Shamir, R. Revealing modularity and organization in the yeast molecular network by integrated analysis of highly heterogeneous genomewide data. Proc. Natl. Acad. Sci., Vol. 101. U. S. A. (2004) 2981-2986
7. Kitano, H.: Biological robustness. Nature Reviews Genetic, Vol. 5 (2004) 826-837
8. Stelling, J., Sauer, U., Szallasi, Z., Doyle, F.J., Doyle, J.: Robustness of Cellular Functions. Cell, Vol. 118 (2004) 675-685
9. Holme, P., Huss, M., Jeong, H.: Subnetwork hierarchies of biochemical pathways. Bioinformatics, Vol. 19 (2003) 532-538
10. Barabasi, A.L., Oltvai, Z.N.: Network biology: Understanding the cells's functional organization. Nature Rev. Genetics, Vol. 5 (2004) 101–113
11. van Dongen, S.: Graph clustering by flow simulation. PhD thesis. University of Utrecht, Center of mathematics and computer science. (2000)
12. Kannan, R., Vampala, S., Vetta, A.: On clustering: good, bad and spectral. In Proceedings of 41st Annual Symposium on Foundations of Computer Science. (2000) 367-378
13. Gaertler, M.: Clustering with spectral methods. Master's thesis. University at Kon-stanz. (2002)
14. Brandes, U., Gaertler, M., Wagner, D.: Experiments on graph clustering algorithms. ESA LNCS, Vol. 2832 (2003) 568–579
15. Guimerà, R., Amaral, L.A.N.: Functional cartography of complex metabolic networks. Nature, Vol. 433 (2005) 895-900
16. Guimerà, R., Amaral, L.A.N.: Cartography of complex networks: modules and universal roles. J. Stat. Mech. Theor. Exp., P02001 (2005) 1-13
17. Eisen, M.B., Spellman, P.T., Brown, P.O., Botstein, D.: Cluster analysis and display of genome-wide expression patterns. Proc. Natl Acad. Sci. USA, Vol. 95 (1998) 14863–14868
18. Quackenbush, J.: Computational analysis of microarray data. Nature Reviews Genetics, Vol. 2 (2001) 418–427
19. Guimerà, R., Sales-Pardo, M., Amaral, L.A.N.: Modularity from fluctuations in random graphs and complex networks. PHYSICAL REVIEW E 70, 025101(R) (2004)
20. http://www.affymetrix.com/support/technical/sample_data/datasets.affx
21. Enright, A.J., Van Dongen, S., Ouzounis, C.A.: An efficient algorithm for large-scale detection of protein families. Nucleic Acids Research, Vol. 30 (2002) 1575-1584
22. Harlow, T.J., Gogarten, J.P., Ragan, M.A.: A hybrid clustering approach to recognition of protein families in 114 microbial genomes. BMC Bioinformatics, Vol. 5 (2004) 45.

Author Index

Lecture Notes in Bioinformatics